DISASTER MITIGATION, PREPAREDNESS AND RESPONSE

THE INTERNATIONAL DECADE FOR NATURAL DISASTER REDUCTION

DISASTER MITIGATION, PREPAREDNESS AND RESPONSE

An Audit of UK Assets

David Sanderson
Ian Davis
John Twigg
Belinda Cowden

The Oxford Centre for Disaster Studies

in association with

Intermediate Technology Publications 1995

Acknowledgements

This project has been made possible by the financial support and technical advice from staff in the Emergency Aid Department of the Overseas Development Administration, and by the support of the UK IDNDR National Co-ordination Committee. The editors have valued the support given by both these organisations.

Grateful acknowledgement is especially given to all those who returned questionnaires and to those who agreed to be interviewed. The editors are also grateful to Roddy Maddox for database development and to Carlos Guerrero for his work during the early stages of the project.

The audit was written, compiled and edited by:

David Sanderson
Dr Ian Davis
Dr John Twigg
Belinda Cowden

The views expressed in this publication are of the editors alone. The editors encourage reproduction of part of all of this audit, but request that in any reproduction the source is acknowledged. Text however may not be modified or adapted, unless specific permission is requested from the editors (c/o OCDS).

© The Oxford Centre for Disaster Studies, 1995

ISBN   1 85339 331 2

The Oxford Centre for Disaster Studies (OCDS)
PO Box 137, Oxford. OX4 1BB, UK.

Tel 01865 202772; Fax 01865 202848

Intermediate Technology Publications
103–105 Southampton Row, London WC1B 4HH, UK

Printed by SRP, Exeter, UK

Contents

Acknowledgements	iv
Foreword	v
Reading the audit	vii
Introduction	viii

The audit	1
Overview	2
• *Organisations*	2
• *Areas of work, specialisms and interests*	3
Regions of activity	8
Hazard expertise	11
Work and skills	14
• *Work content*	14
• *Skills*	17
Education and training courses	20
• *UK based academic courses*	20
• *UK based training courses*	22
Networks	24
• *Questionnaire results*	24
• *Professional/technical associations*	27
• *Development networks*	27
• *Conclusions: problems and potentials*	29
Funding	32
• *The Overseas Development Administration*	32
• *Other official sources*	36
• *Charitable trusts and foundations*	37
• *Other funders*	39
• *Conclusions*	45
Information sources	48
• *Libraries and information centre*	48
• *Bookshops*	48
• *Journals*	49
• *Databases*	49
• *The Internet*	50
Summary of organisational activities	53
Current activities	58

The directory	75
Directory of organisations	76
IDNDR focal points	101
Directory of individuals	111

Appendices	129
1. World Conference on Natural Disaster Reduction, Yokohama, 1994: *Strategy Paper*	130
2. Current networks	132
3. Summary of research methodology and audit questionnaire	135

Charts

1. Proportional relationship of organisation activity according to region   8
2. Three main regions of activity according to organisations   10
3. Breakdown of organisational activity according to hazard   11
4. Percentage of work content according to organisation   14
5. Four key areas of work content according to organisational activity   16
6. Five highest scoring skill areas according to organisational activity   17

Tables

1. Percentage activity of organisations according to region   8
2. Percentage organisational activity of according to hazard   11
3. Percentage activity of organisations according to work content   15
4. Percentage of skills according to organisation   19

Caption Boxes

Reading the Audit   vii
British Red Cross, International Division   10
Oxfam Emergency Aid Department   13
Ove Arup Engineers   16
Rob Stephenson, independent consultant   18
Oxford Brookes University; complex emergencies and humanitarian aid   20
The London School of Hygiene and Tropical Medicine   22
Society for Earthquakes and Engineering Dynamics (SECED)   25
Emergency Aid Department, ODA   33

Foreword

The aim of the audit was to gain a coherent overview of the current UK assets available in the field of Disaster Mitigation and Preparedness (DMP). Specifically, to:

- identify strengths, weaknesses and gaps in UK DMP capability;

- encourage and enhance networking and skill-sharing within the DMP community;

- contribute to a more comprehensive and co-ordinated response to international disaster needs.

To these ends the information gathered and presented in this document represents the findings of over 170 returned completed questionnaires from charities and non-governmental organisations (NGOs), academic departments, government departments, consultancies, consultants and private companies. Additional information and comment has been added by the editors and members of the UK disaster mitigation and preparedness (DMP) community, including those from the UK Co-ordinating Committee of the International Decade for Natural Disaster Reduction (IDNDR), 1990-1999.

The text is divided into two sections: *the audit* and *the directory*. The former contains the analysis, recommendations and findings of the research, whilst the directory is a compilation of gathered information, presented in usable form for the DMP community.

The audit

The audit comprises the following:

- Overview
  The overview contains a summary of observations, conclusions and recommendations resulting from the research;

- Activities findings
  Findings for each of the sections identified under *activities* in the audit, presented as statistical information with comment and lists of useful information. There are seven sections: *Regions of Activity, Hazard Expertise, Work Content and Skills, Education and Training Courses, Networks, Funding* and *Information Sources*.

  The first four sections include two sets of statistics: a pie chart giving the breakdown of organisations according to activity, for example 8 per cent of all organisations responding to the audit are involved in landslide; and a table giving the percentage of activity of a particular organisation according to activity, for example 37 per cent of individual consultants replying to the questionnaires stated work in famine. The final three sections include listings, addresses and telephone numbers of relevant organisations;

- Summary of organisational activity
  This section summarises in tabular form the activities of organisations according to region, hazard type and skills;

- Current activities
  Current activities presents information relating to current projects, including organisation/ individual and contact name, project title, objectives and duration.

The audit also contains caption boxes: brief overviews of organisations and individuals, resulting from interviews, intended to present a fuller picture of current activities in this field.

The directory

The directory comprises the following:

- Directory of organisations
  Information presented of all organisations returning questionnaires includes name, address, contact name and position, income, expenditure, fax, phone and E-mail address, number of staff and mission statement;

- Directory of individuals
  This includes the name and contact address of every individual in the audit, arranged as address label format for best usage;

- IDNDR focal points
  gathered information of key UK and international individuals available for comment and broad discussion of their areas of expertise, which make up seven sections: *General Knowledge of Disaster Preparedness/Mitigation, Hazard Types and Related Sectors, Sectors, Country Knowledge, Government focal points, NGO focal points* and *International focal points.* Focal points have been ratified by the UK IDNDR Co-ordinating Committee and the individuals themselves.

The appendices

There are three appendices, comprising:

- The Strategy Paper that emerged from the World Conference on Natural Disaster Reduction, Yokohama, 1994;

- Current networks: an unedited listing of all the networks referred to in the returned questionnaires;

- Research methodology: a summary of the research carried out to produce this audit. Also included is a copy of the Audit Questionnaire.

Reading the audit

Before reading the audit it is important to make the following points:

- The editors and sponsors of this exercise were encouraged by the response to the questionnaires, and the willingness of individuals to agree to becoming focal points. However, there are some gaps: some individuals and organisations failed to submit their response in time for inclusion in the analysis (although they are included in the directory). The editors also may not have been able to contact all who should have been included in this exercise. It is hoped that later editions will cover such omissions;

- The information contained in this audit reflects the responses given on the returned questionnaires: it was felt *not* to be the role of the editors to alter any returned information (unless obviously incorrect) since it would have been impossible in practice to check all incoming information. This is a particularly important point when reading the statistical findings, which are based on the assumption that answers given by returned questionnaires are truthful and correct. Hence, although all findings are accompanied by comment, it is nevertheless important to measure the statistical finding with a critical interpretation based on the reader's own knowledge;

- From the earliest stages of the project it was felt important to gather information relating to the currently most pressing activity of many of those in the audit, namely complex emergency and refugee activities. Although outside the scope of natural hazards this information is included;

- The scope of the research has been broadened from the original aim of gathering mitigation and preparedness information only. This has been carried out in recognition of the different understanding of these terms by the wider community, and of the difficulty in identifying these specific activities within programmes. Hence the editors have made the audit inclusive rather than exclusive in its content, in order to contain much of the valuable information received, which would have been discarded otherwise;

- Finally, as a first exercise in this field, the aim was to investigate *breadth* rather than *depth*. Some therefore may feel that more detail would have been useful. Where this is the case it is hoped the audit will prove useful in providing a platform for more detailed research by others.

Introduction

Before undertaking this audit, the general assumption of the editors, and to those spoken to in association with the project, was of a series of strengths and weaknesses in the field of the UK capacity to prepare against or mitigate future disasters. The findings from the audit now identify these strengths and weaknesses, as well as providing information on the activities of organisations involved in disaster mitigation, including consultants and consultancies, NGOs, academic bodies, private companies and governmental departments.

Whilst the UK is largely free from major hazards, there is nevertheless extensive work proceeding in this field involving British organisations and individuals working in other countries, mostly in the Southern hemisphere: Africa, Asia and Latin America. Work of this nature in these regions has no doubt been assisted by extensive post-colonial contacts, the extensive development emphasis in UK academic institutions and the tradition of humanitarian voluntary aid which has been particularly strong in the UK.

From the audit, the major UK contributions would appear broadly to lie in the following areas:

- The prediction, monitoring and management of drought;

- The development of food security systems in drought-prone areas;

- Seismology and engineering seismology, strongly represented in British academic and consultancy bodies (this focus probably derives from the primary development of the subject in the UK);

- The development of cyclone-warning systems (The Meteorological Office continues to fulfil an international role in the global cyclone-warning network);

- The systematic approach to disaster management developed by British NGOs, which is well documented and widely adopted internationally. This work has been a collective effort of such groups as Oxfam, Save the Children Fund, The British Red Cross and Action Aid, amongst others;

- UK publishing that continues to make a significant contribution through key journals (the primary one being *Disasters*) which provides a vital dissemination tool for research findings. In addition publishers such as Intermediate Technology Publications, Wiley and Oxfam Publications have maintained a steady flow of materials on this theme. Publishing remains the key channel for the development of knowledge (greatly enhanced by the growth in use of the Internet);

- Disaster Management Training, a very extensive and growing sector; for example the United Nations Disaster Management Training Programme (DMTP) has been extensively supported by UK expertise;

- Hazard-resistant low-cost building construction, which has been a major subject in this field since early programmes in the late 1970's with key texts being produced and disseminated.

The IDNDR

This audit appears midway through the International Decade for Natural Disaster Reduction (IDNDR). It allows us to judge, in part, how well equipped the UK disaster community is to further the objectives of the IDNDR. Readers can make their own assessments from the data and commentaries that appear in the following pages, setting this information against the IDNDR's 'Strategy for the Year 2000 and beyond' issued at the World Conference in Yokohama in 1994 (see Appendix One for the Conference Strategy Report).

However the audit is designed to go further than this. It is intended to stimulate action: to contribute directly to the IDNDR's global strategy. Perhaps this appears an ambitious aim; but we hope the audit will make a discernible contribution in four of the strategic areas highlighted at Yokohama.

The IDNDR seeks to develop a global strategy of prevention (point A of the 'Strategy to the Year 200 and beyond'). As an analysis of resources available for disaster prevention, the audit is part of this process. Its national-level findings make up a tiny part of the picture world-wide but can be set against similar exercises being planned or implemented elsewhere, or against regional surveys such as the directory recently compiled by *La Red* for Latin America. There is a particular need for analysis of capacity in countries of the South.

The strategy calls for the identification and networking of existing centres of excellence (point E). The audit identifies skills and capacities in the UK, which is a prerequisite for effective networking. Moreover, it sets out some suggestions for better networking among the disaster community. Linked to this is the need for improved co-ordination and co-operation in research and for more interdisciplinary research (point K).

However, whilst studies such as this can stimulate co-operation, they do not guarantee it. There may be institutional or personal obstacles. In the UK, where competition for relatively limited research funding is intense, financial barriers may be the most formidable.

The Yokohama conference called for higher priority to be given to the compilation and exchange of information on natural disaster reduction (point M). We trust that the results presented by the audit will be valuable in supporting this process. The very commissioning of the audit reflects the importance attached to this work by the UK's IDNDR Committee and the Overseas Development Administration (ODA).

However, the audit is a starting point, not a conclusion. Its findings must be taken up and acted upon by organisations and individuals in the UK, and perhaps beyond, if it is to make any significant impact on the progress of the IDNDR.

The audit

Overview
Regions of activity
Hazard expertise
Work and skills
Education and training courses
Networks
Funding
Information sources
Summary of organisational activities
Current activities

Overview

The following overview of the key findings of the audit is organised into two key headings: Organisations and Areas of Work, Specialisms and Interests.

Organisations

The questionnaire used seven categories to cover organisations represented in the UK disaster community: charity/NGO, private company, academic/research body, consultancy, individual consultant, government department and intergovernmental agency. Such division into categories is useful in understanding the community as a whole but in real life distinctions are less clear: a large, diverse and multi-disciplinary community cannot be captured neatly within a series of rigid compartments.

In reality it is possible for a single organisation or individual to fit into more than one category. For example, an NGO or academic institution may also provide a consultancy service in some shape or form (indeed, may have to in order to survive financially); an individual consultant may also be involved as a regular researcher with another agency or with an academic link. The division between consultancy firms and private companies is particularly fluid. Replies to the questionnaire demonstrate this issue clearly, with several respondents ticking more than one box. Cambridge Architectural Research (CAR) is a case in point: it identified itself as an academic/research body, consultancy and private company.

Nonetheless, the figures are revealing. They show the disaster community to be split broadly into four groups: the voluntary sector (charities and NGOs), academic and research institutions, the 'commercial' sector, and the state sector. In numerical terms the first three categories account for roughly a third of respondents each. The fourth group is influential but small in number. Looking within the categories we find a rich diversity:

Charities and NGOs

The range of agencies here mirrors the variety within the voluntary sector generally. They range from large development and relief organisations with annual budgets running into millions of pounds, for example Oxfam and Christian Aid, to much smaller agencies focusing on single issues or locations.

Academic and research institutions

A wide range of academic organisations and departments are involved with disasters. They include a diversity of disciplines, including architects, planners, economists, environmental scientists, health experts, geographers, civil engineers, anthropologists, nutritionists and ecologists. Organisations include centres for regional studies, specialist units focusing on individual hazards such as earthquakes or floods, and development studies departments.

The 'commercial' sector: private companies

The small number of entries under the heading *private company* was surprising although this may indicate that commercial operations attach relatively little importance to this kind of survey and therefore did not reply to the questionnaire. Insurance companies were most likely to appear in this category.

University departments were prominent among the *consultants*, making up well over a third of respondents in this category. The remainder largely comprised commercial firms and a small number of individuals. Most of those who marked themselves as *individual consultants* were fully freelance although several were linked to academic departments.

The state sector

The group of *government departments* included sections of two government ministries (the ODA and Department of the Environment) and other national bodies (the Meteorological Office and National Rivers Authority). However, it is recognised that there are additional government ministries and agencies with an involvement in this field such as the Building Research Establishment.

There were two entries under the category *intergovernmental agency*. One was the Natural Resources Institute, which might equally well be deemed a government department and is in any case being privatised in effect. The second was the UK Committee for UNICEF.

Only one organisation, the Crown Agents, described itself as a *public corporation* (a category not used in the questionnaire).

Areas of work, specialisms and interests

This data can be understood best from the accompanying tables, to be found in each of the audit sections, which plot types of organisations against links with other agencies, work in different geographical regions, expertise in hazards, the nature of their activities and the skills they possess. A limited commentary is provided here to highlight some of the main features revealed in the figures.

Regions

The audit identified ten global regions. The region most recorded for activity was unsurprisingly Africa followed by South\South East Asia. Eastern Europe and the former Soviet Union countries showed up as an important area for UK NGOs, consultancies and government departments (reflecting the current emphasis in international aid policy and hence in funding priorities). In contrast, the regions most 'neglected' were The Caribbean (traditionally a strong area of activity for the UK), Latin America including Mexico, East Asia (China) and the South Pacific.

For Africa, disasters addressed were recorded in order of priority as being:

1. Drought (especially in South Eastern Africa)
2. Complex Emergency (mostly in Rwanda, the scene of enormous activity, especially by NGOs and intergovernmental agencies)
3. Famine[1]
4. Flood
5. Disease and Epidemic

Most organisational types (NGO, consultancy, etc) followed this pattern.

Commercial and consultancy interest was more pronounced in the more developed regions such as Australia and the Pacific, and the USA and Canada. Here other categories were not greatly involved. However, Western Europe appeared to be significant for several types of respondent. Latin America was of average significance for most categories, yet there was a reasonable level of involvement in the region by all except individual consultants. It is interesting to speculate on the likely picture in that region in a few years time, since local capacity and regional co-operation are growing rapidly.

Such a breakdown of priorities of Africa and Asia as the top two areas might be attributed to colonial ties/influence. Of the 'neglected' regions (by the UK), Latin America including Mexico has traditionally been the domain of the USA for aid. The shift in activity away from The Caribbean could be attributed to global shifts in need and aid distribution over the last decade, for example to Africa, Former Soviet Union, Eastern Europe.

---

[1] A distinction has been made between famine and drought, since famine is usually a consequence of a series of events of which drought may be only one factor.

Hazards

The returns show that UK organisations and individuals are involved in dealing with a wide range of hazards. Interest and expertise are well distributed, though the influence of the commercial categories is more pronounced in 'technical' hazards such as floods and earthquakes while NGOs and academics are mostly involved with drought and famine.

Complex emergencies dominate the figures, with all categories being extensively involved. For NGOs/charities, academic/research bodies, consultancies and individual consultants this was the single most important area of current work.

Work and skills

The audit questionnaire identified two overlapping and complementary groupings: *Work Content* and *Skills*.

Work Content

Of the thirteen identified areas of work content, the four key categories of work being undertaken in order of reported activity were:

1. Risk assessment
2. Relief and humanitarian aid
3. Vulnerability assessment
4. Community level disaster preparedness

Conversely the lowest levels of involvement were in major engineering, structural mitigation measures and public awareness raising (North and South). Also it appears few organisations are carrying a torch for gender issues![2]

For NGOs and other voluntary sector organisations, relief work and community-level preparedness were the main areas of interest, with national-level planning and engineering work among the lowest priorities. NGOs still attach relatively little importance to raising public awareness of disaster protection opportunities even though this is surely a prerequisite of promoting disaster mitigation and effecting changes in policies.

As a whole, the commercial group of companies, consultancies and individual consultants were most interested in risk and vulnerability assessment. Among private companies national-level preparedness planning was relatively significant, while for consultancies of all kinds relief and community-level preparedness were more important. Engineering and structural mitigation measures were of greater interest among this group than in other categories, but not to a vast extent. Academics had similar priorities: risk and vulnerability assessment, and community-level preparedness.

Among government and intergovernmental agencies warning systems assumed much greater significance and, with the others, risk assessment was a high priority[3]. Community-level work was not prominent here. Conflict resolution is perhaps going to be the fastest 'growth industry' in disaster work in the next few years. Most categories of respondents appeared moderately interested but only consultants (groups and individuals) appeared to be particularly involved as yet.[4]

However, it appears that expertise and experience, if it is as substantial as indicated by the findings, is not being documented to reflect the level of activity. If this is the case, then greater attention to

---

[2] NGOs and academic organisations, who one would expect to be more concerned about this area, were more heavily involved than others but even so the level of involvement was low within the groups as a whole.

[3] The high number of organisations and individuals that have cited 'risk assessment' as a key area of work is rather perplexing since the literature available on this topic, particularly vulnerability assessment, is scarce. If risk assessment has a high profile, then this would presumably be reflected in current writing/conferences

[4] This finding was also reflected in the replies to the question on skills (see the following section).

documentation and dissemination of experiences is needed. Such a request is in particular addressed to the NGO sector which, often at the forefront in grassroots initiatives, does not always disseminate its findings, even though internal reports are produced.

Skills

The questionnaire identified 26 'skill areas', ranging from forestry, training, physical planning and remote sensing to volcanology, geomorphology and anthropology. The clearest finding was the stated prominence of training in every responding group, confirming the UK disaster community's extensive commitment to this activity.

The commercial group identified in 'skills' was particularly active in what can broadly be termed technical specialisms: research (technical and social science), building and architecture, engineering, energy, insurance and physical planning. In general their expertise was well spread but least in the specialist scientific areas such as volcanology, seismology and meteorology, which remain the province of academics and researchers.

Individual consultants were the most active group in food security work which, with health and training, were the areas where the NGO sector was best equipped. Activity in agriculture and forestry was also most marked among NGOs. Academics' main skills included those most relevant to wider development issues: food security, health and nutrition, agriculture, anthropology, conflict, indigenous knowledge and appropriate technologies, and economics. NGOs and consultancies/consultants recorded particularly high responses to training, possibly indicating a high commitment to skills transfer. Training included UK-based courses (of which there are a variety in content and length, see below) and in-country courses, in which many organisations were involved.

What was surprising however was the very low levels of NGO activity in research: 15 per cent for both technical and social science research. In contrast 46 per cent of consultancies claim to carry out research in these areas.

The lowest recorded responses of listed skills were in forestry, meteorology and energy. NGOs reported the highest activity in forestry, whilst private companies and consultants were highest in energy; government departments were highest for meteorology.

One finding regarded a bias of NGOs towards rural areas: NGOs reported a low level of activity in built environment, yet much higher in agriculture, forestry, etc. It may be extrapolated, in tandem with anecdotal knowledge and reports of current activity, that NGOs' work is mostly in rural areas, and that they have been slow to develop strong urban-based work. This split may suggest the skills and training of UK NGO staff or the rural bias of NGOs' southern partners, but it could also be argued that rural areas contain the worst poverty (as well, of course, that the majority of people still live in rural areas). There are however strong developing pockets of urban expertise (for example IIED and Homeless International); also some larger NGOs are claiming an increasing focus on urban need.

Links

The analysis shows that UK-based organisations and individuals have their strongest links with international and national NGOs, government departments and academic/research institutions; while the weakest links across the board are with private companies, individual consultants and networks.[5]

Most categories of respondent were, not surprisingly, likely to work most closely or frequently with agencies of their own kind. NGOs' strongest links were with other international, national and grassroots NGOs, and it seems that grassroots organisations depend particularly on external NGOs for their contacts with expertise in other countries. The main exceptions to this 'like with like' emphasis were in the commercial group: companies did liaise with other firms but consultants, individual and collective, placed other consultants and private companies low on their list. It may be speculated that professional rivalry is a factor here, with small consultancy operations feeling particularly threatened by competition.

---

[5] Regional networks are relatively new and this may explain the limited contact. NGOs are best linked to them, followed by academic and research organisations.

Networks/information technology/information sources

The audit found that, for the majority of questionnaire respondents, the use of formal networks is neither widespread nor seems to be of great importance.

The formal networks most frequently cited were:

- Specific disaster networks, such as the Relief and Rehabilitation Network (RRN) of the Overseas Development Institute (ODI)
- Development networks, for example the Development Studies Association (DSA)
- Professional associations, such as the Institution of Civil Engineers (ICE) - and the associated body of SECED
- Geographical interest groups, for example the European Network of Bangladesh Studies

However from anecdotal evidence it is clear that, in such a relatively small community, informal networks are essential in maintaining information flow, knowledge of current activities, 'who is doing what', etc. Networking, a critical need for rapid information sharing, could be enhanced by adoption of the following:

- Information sharing between individuals, organisations and professions
- The setting up of meetings and the development of partnerships
- The creation of institutional focal points, (i.e. an *institutional base* for networks)

The IDNDR National Co-ordinating Committee and its working groups are already fulfilling a role in promoting the growth of networking, and there is potential here to take this further.

The Internet

The Internet is the fastest growing form of national and international communication, information exchange and networking. Already there exist 'home pages' for the IDNDR, as well as disaster 'discussion groups' in most if not all hazard types.

Resources

Libraries

Information and knowledge is located in various locations, both accessible and private. Of the latter, consultancies and private companies may have built up substantial bodies of knowledge which remain inaccessible to researchers; of the former, universities/academic institutions such as the Institute of Development Studies (IDS) in Brighton or the Refugee Studies Programme (RSP) in Oxford are more accessible. There is currently, however, no single central body organising literature mitigation/ preparedness regarding disaster expertise or information. The development of such a centre would be of great benefit to the DMP community.

Academic courses

The audit gathered information only on those development courses (of which there are a number) which offered hazard-related options as part of a course. The findings were that the majority of course were pitched at graduate and postgraduate (MSc, MPhil and PhD) level; undergraduate courses components often featured as part of geography or related degrees. A new course beginning in September 1995 is a BSc (Honours) degree course in International Disaster Engineering and Management offered by The Fire Service College at Coventry University. Of the course components offered, either in training or academia, the editors could find no comprehensive dedicated list. Hence the list compiled in this audit is a contribution to the assembling of such knowledge.

Although there have been recent new course components developed, (for example the complex emergency option as part of the MSc in Development Practices offered at Oxford Brookes University) there are no courses offered with hazard studies as the key focus.

Funding

The audit found that, apart from money made available as a result of humanitarian appeals, eg Rwanda, there are very few available sources of funds or dedicated budget lines for funding bodies. Two principle sources that do exist include the Emergency Aid Department of the Overseas Development Administration and the European Community Humanitarian Office (ECHO) of the Economic Union.

The ODA is by far the largest source of funds in the UK for all activities related to disasters. A significant feature of the British Government's aid programme, and of official development assistance generally, is the increasing proportion of humanitarian aid in the total. The same trend is visible within the ODA's country programmes. The European Union gives massive amounts of humanitarian aid[6]: nearly ECU 605 million (£465 million @ £1 = ECU1.3) in 1993, and over ECU760 million in 1994. Its funding in this area has risen seven-fold in the last four years. Funds are contributed by member states.

Charitable trusts formed the bulk of the funding organisations sent questionnaires by the audit team (160 specific funding questionnaires were sent). They were selected because their directory entries indicated an interest in disasters. Hardly any of these were prepared to divulge details of their work and it is likely that they are interested in relief rather than preparedness and mitigation.

NGOs and funding

Some operational NGOs are also grant makers, the most significant such as Oxfam and Christian Aid being multi-million pound organisations which run their own emergency projects, support local NGOs, and may need consultants for technical assistance, studies and evaluations. Other NGOs, on a smaller scale, have similar aims and act in similar ways. Corporate giving in the UK is on the increase although the levels of funding and strategic planning have a long way to go to catch up with practice in the United States.

The relative difficulty of assuring long-term *consistent* funding, especially for NGOs undertaking relief programmes, leads to a recommendation for the education of funders in disaster response, ie to promote an awareness of the need to support long term recovery, thus reducing risks of disaster recurrence. Linked with this is the need to create long term partnerships and collaborations.

---

[6] Its definition of humanitarian aid comprises food aid, emergency aid and aid to refugees.

Regions of activity

The audit questionnaire identified ten regions of activity: Africa, the Middle East, East Asia (including China), South/South East Asia (including India and Bangladesh), Eastern Europe/former Soviet Union, Western Europe (including former Yugoslavia), the Caribbean, Australia and the Pacific, Latin America including Mexico, and the USA and Canada.

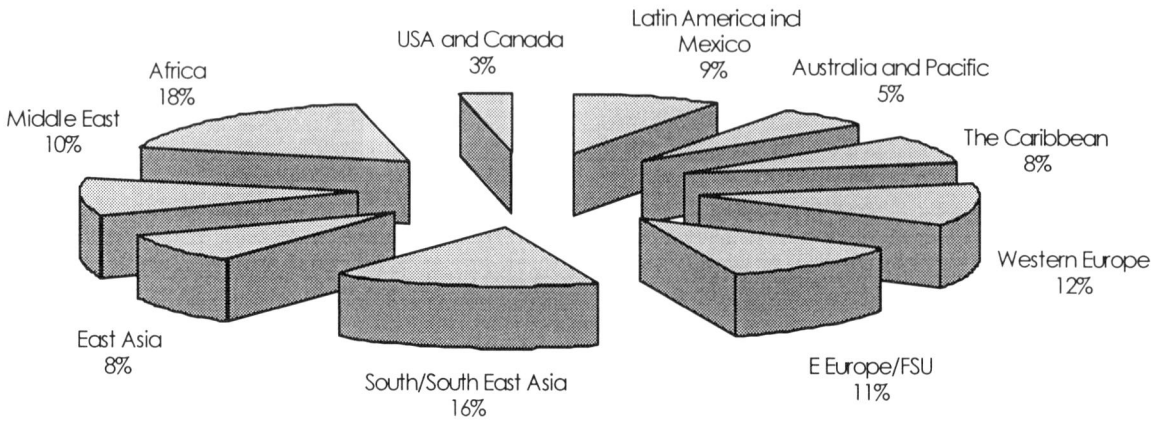

Chart One Proportional relationship of organisational activity according to region

The above pie chart indicates the proportional relationships of organisational activity according to region. Hence, of all the questionnaires returned, 18 per cent indicated activity in Africa, the highest region, followed closely by South/South East Asia (16 per cent). Western Europe[7] is the third with 12 per cent followed by Eastern Europe/Former Soviet Union with 11per cent. The Middle East is the fifth highest area with 10 per cent.

	Latin America inc Mexico	Australia and Pacific	The Caribbean	Western Europe	E Europe/FSU	South/South East Asia	East Asia	Middle East	Africa	USA and Canada
Charity/NGO	29	13	21	23	44	54	27	35	65	6
Private Company	31	23	23	38	23	46	31	23	46	15
Academic/research body	34	13	21	40	32	42	21	30	57	13
Consultancy	36	29	29	36	50	68	46	46	75	18
Individual Consultant	11	5	21	37	26	53	21	53	63	11
Government Department	43	29	43	71	57	71	29	29	71	0

Table One Percentage activity of organisations according to region

---

[7] The high percentage of activity for Western Europe is assumed to be accounted for by respondents including Former Yugoslavia in this region.

Table One indicates as a percentage the activities of specific organisations according to identified region. Hence it can be seen that, of the data collected, roughly 2/3 of NGOs (65 per cent) were active in Africa, whilst just over half (54 per cent) were active in South/South East Asia. However the third key area of activity for NGOs was Eastern Europe/Former Soviet Union at 44 per cent, just over double the activity of the Caribbean at 21 per cent. Similarly whilst for the overall breakdown Latin America including Mexico accounted for only 9 per cent of overall activity, 43 per cent of Government Departments were active in that region (the key actor for all organisational types). To view Africa as a case in point, it can be seen from Table One that three quarters of all consultancies registered activities in Africa, followed by government departments (71 per cent), charities/NGOs (65 per cent) and individual consultants (63 per cent). According to returned questionnaires, the five main hazard types addressed in Africa were:

1. Drought (35 per cent)
2. Complex emergency (34 per cent)
3. Famine (34 per cent)
4. Flood (25 per cent)
5. Disease and epidemic (22 per cent)

Of drought, the main hazard type addressed, 40 per cent of NGOs registered activity, compared to 31 per cent of academic/research bodies and 37 per cent of consultancies. Examples of current projects by academic/research bodies include the development of recommendations for drought response in Kenya by the Food Studies Group and disaster management training throughout southern Africa for UNDP/DHA by the Disaster Preparedness Centre at Cranfield University; whilst NGO activities include a series of drought/food security training of training workshops in southern Africa by Tear Fund, funded by ODA and ECHO. In contrast to drought, over half (56 per cent) of the consultancies returning questionnaires stated their involvement in complex emergency issues in Africa, compared to only 25 per cent of academic/research bodies and 37 per cent of charities/NGOs.

Most current complex emergency activities of NGOs are focused on Rwanda, for example relief and rehabilitation from Christian Aid, a major review of the relief effort on behalf of ODA by the Overseas Development Institute (ODI), healthcare and rehabilitation (medical supplies, restoration and health structures training) by MERLIN and emergency relief to displaced communities by ACORD. A comprehensive overview of the Rwandan crisis, *Rwanda: Dilemmas of a Total Disaster* is provided in the *World Disasters Report, 1995*; International Federation of Red Cross and Red Crescent Societies, Geneva, 1995. Pages 59-68.

Organisations and regions of activity

Chart two on page 10 indicates the activities of organisations in the three highest rated regions of Africa, South/South East Asia and Western Europe. Therefore for instance it can be seen that approximately 70 per cent of government departments who returned questionnaires stated activity in Western Europe, South/South East Asia and Africa.

Academic/research bodies stated high percentages of activity in both Africa and South/South East Asia. For instance the ODI, as well as carrying out consultancy work in Rwanda (see above) is currently carrying out research into the economic and financial aspects of drought on sub-Saharan African economies for ODA/World Bank, and the economic impact of natural disasters in South East Asia and the Pacific.

It can be seen that whilst all rated highly for Africa, charities/NGOs registered a lower level of activity for Western Europe. This may reflect the relative recent of conflict in Western Europe compared to the long term involvement of NGOs in Africa. Also, the lowest region of activity for consultancies was Western Europe, whilst private companies ranked South/South East Asia and Africa the same for intensity of activity.

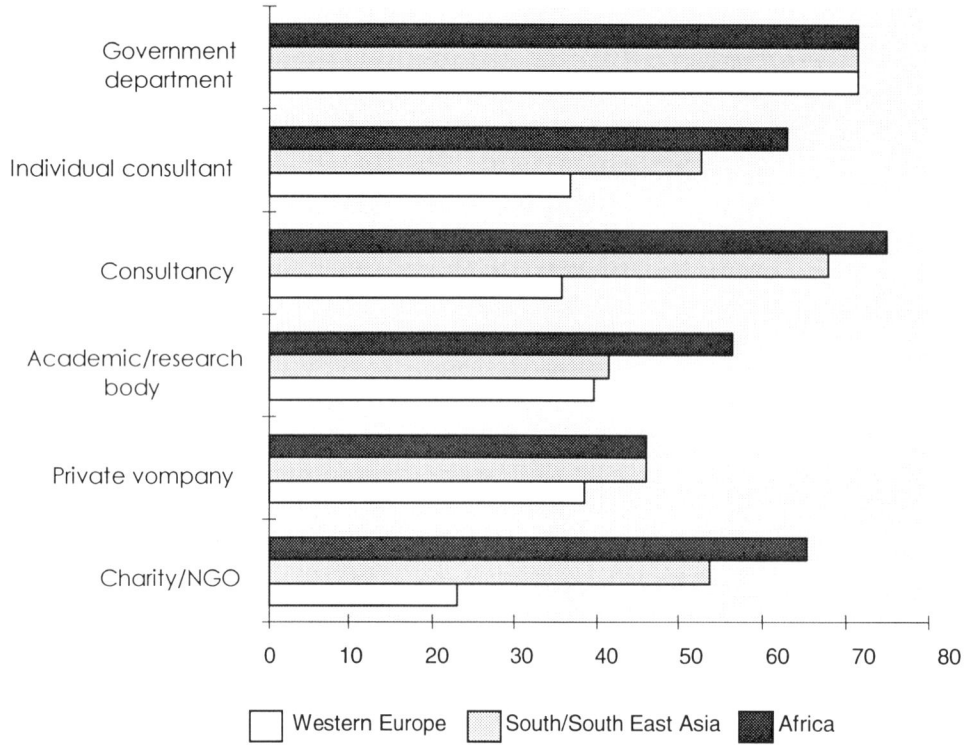

Chart Two Three main regions of activity according to organisations

British Red Cross, International Division

The British Red Cross has been working in emergency relief since 1870. The International Division of the British Red Cross currently has 97 delegates working overseas in a variety of relief and development roles with local Red Cross Societies. In 1994 they responded to over 40 major emergency operations, of which the largest were in and around Rwanda and the former Yugoslavia.

Mike Adamson, the Head of the International Development Department, believes that the strength of the International Red Cross movement lies in its network of local Red Cross Societies: 'In almost every country in the world we have a local partner and we work through them. This is what makes the Red Cross special. In essence the Red Cross is an international network of local emergency organisations. Volunteers at the community level and at branch level address the silent and day-to-day emergencies that never catch the headlines as well as being in a better position to respond to larger emergencies when they occur. The Red Cross's long term work is about improving emergency preparedness and reducing vulnerability to risk and hazards. This is achieved through a range of measures: from emergency shelter in Bangladesh to community health projects and water programmes in Ethiopia. The institutional development of local Red Cross Societies through management training and organisational development is also contributing to emergency preparedness, a key activity.'

The British Red Cross is facing a dilemma similar to other organisations: how to protect longer term projects like disaster prevention and mitigation in the face of rapid onset, large scale emergency response needs. Mike Adamson states, 'we are developing a portfolio of longer term programmes which contribute to disaster preparedness and mitigation while also bringing a development philosophy into our emergency work so that programme beneficiaries and our local partners come out of an emergency operation stronger than before it happened.'

Hazard expertise

The audit identified nine major hazard types. These are represented in chart three below:

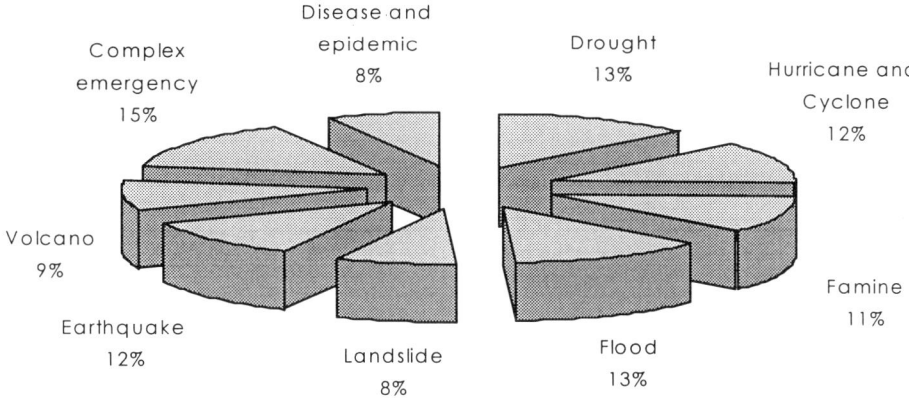

Chart Three Breakdown of organisational activity according to hazard

	Drought	Hurricane and Cyclone	Famine	Flood	Landslide	Earthquake	Volcano	Complex emergency	Disease and epidemic
Charity/NGO	40	40	42	37	19	33	21	48	33
Private Company	23	54	23	38	46	62	38	38	8
Academic/research body	34	21	28	17	9	17	23	40	17
Consultancy	43	43	39	46	21	39	21	57	29
Individual Consultant	42	37	37	42	32	32	21	63	32
Government Department	71	43	43	71	29	43	43	43	29

Table Two Percentage of organisational activity according to hazard

From chart three and table two above it can be seen that, whilst there is a significant emphasis on certain hazard categories (often termed slow-onset disasters), there is a surprisingly even spread of hazard involvement across all the organisations that returned questionnaires.

It can be seen that major concentration of effort is being extended to drought, famine and complex emergencies, whilst less involvement is being devoted to landslides and volcanic hazards. This focus reflects both the frequency of hazard events, their social, economic and political impact and their relative importance to such fast impact disasters as landslides and volcanoes in terms of 'loss of life potential'.

In the recently published *World Disaster Report*[8] statistics are quoted that in 1993, 250 000 people were killed in war, 17 million fled as refugees and up to 26 million were displaced within their own countries. In

---

[8] Cater N. & Walker P. (eds) *The World Disasters Report 1995* International Federation of Red Cross & Red Crescent Societies, IFRCS, Geneva, 1995.

contrast the most powerful volcanic eruption of the past fifty years (Mount Pinatubo in The Philippines) has accounted for approximately 450 deaths.

Thus the UK response to hazards is largely driven by need. It may also be stated that an additional factor is a combination of geography and history in that most of the major drought/famine/ complex emergencies of recent years have been in Africa where the UK has such extensive post-colonial links.

Inevitably the respondents to the questionnaire have provided answers which cover both their involvement in hazards in pre and post disaster contexts since it is difficult, if not impossible, to separate such matters. Therefore some of the concerns noted may not be specifically related to preventive activity.

General comments

The NGO response is rather deceptive: whilst they report involvement in disaster mitigation/ preparedness across all hazard categories, the audit did not attempt to quantify the extent of this commitment. It is likely that if this audit had been undertaken a decade earlier there would have been a greater proportion of NGO effort in preparing and mitigating fast impact disaster events such as floods, volcanic eruptions and earthquakes. The emphasis now is clearly on large-scale relief efforts in the areas of drought and complex emergency.

A further encouraging trend has been the growth of national *self sufficiency*, where countries are managing their own disasters without the need for large scale international appeals for aid. Examples include India, The Philippines, China and Colombia. This increasing self reliance particularly applies to 'localised' disasters such as earthquakes and landslides.

Current priorities for NGOs include food security work by Oxfam's partner organisations in the Indian subcontinent and South East Asia and cyclone warning systems and cyclone shelters by Save the Children Fund in Bangladesh. In contrast the widespread involvement in food security/early warning systems for drought have been given attention by most major NGOs.

However the issue of whether or not NGOs have been working in disaster mitigation is partly a question of definition. The *Oxfam Handbook for Development and Relief* states that: 'The best form of disaster mitigation is through equitable social and economic development, that builds on people's strengths and tackles the causes of their vulnerability' (page 835). In this sense, most of Oxfam's work could be described as disaster mitigation. Thus, if all disaster mitigation is seen as occurring under a development umbrella then most of the development NGOs are active in this field.

It may be worth noting here that the frequently stated view that all development work inevitably covers mitigation is not accurate. Whilst it is broadly true that vulnerability to disasters relates to poverty, and therefore as poverty is reduced, exposure to risk will diminish, nevertheless, this argument does not address the fact that mitigation measures can be highly specific and require sustained attention from NGOs as well as governments. For example, California, one of the richest places on earth, gives detailed attention to disaster preparedness and mitigation, and this is one of the primary reasons why recent disasters have caused so few casualties.

It is hoped therefore that NGOs will progressively develop policy statements, technical expertise, and dedicated funds to ensure that preparedness and mitigation measures take place in the hazard prone countries in which they work. Projects of course also need to be sustainable, ie that they do no get wiped out in future disasters.

Disasters have been described aptly as 'unsolved development problems'. Therefore it is of critical importance to regard disaster problems as well as intervention within a developmental rather than relief culture. One of the achievements of the IDNDR has been to reinforce this concern.

The lack of involvement in certain fast-impact disasters is *not* due to declining need; it is more a reflection of three factors:

1. Diversion of resources to cope with unprecedented demands in Africa, Eastern Europe and the former Soviet Union;

2. A lack of disaster preparedness/mitigation figuring in NGO funding policy statements, hence their absence from projected budget lines;

3. A lack of awareness that preparedness/mitigation is part and parcel of responsible development in hazard prone areas.

The spread of work by consultant organisations is remarkably even, reflecting the diversity of activity within the various sectors. It is also apparent that consultant activity is similar for both organisations and individuals relative to various hazard types.

One area of British consultancy that is particularly strong in this field concerns the extensive work by civil engineering consultants on a global basis in the areas of flood prevention, earthquake resistant building construction, soil stabilisation relative to landslide risk etc. For example the most extensive flood mitigation programme in the world, the Bangladesh Flood Action Plan (FAP) continues to receive major inputs from British consultants which is totally disproportionate to the minor scale of our domestic flood problem.

The academic research community is extremely active in the drought/famine/complex emergency field with attention to hurricanes and volcanoes but with reduced concern for flood, earthquake, disease, epidemic and landslide hazards. It is possible that this priority range reflects the academic emphasis within the UK field of development studies. The UK has an international reputation for academic studies in this field and many of these centres have a strong emphasis on rural development/food security issues. It is surprising however that volcanic hazards attract the level of attention within British universities relative to a reduced concern for say floods which is a far more significant hazard than volcanoes in terms of loss of life and damage to property. It would be useful for further study in the academic/research community in risk reduction to determine the ratio of activity to basic and applied research.

Oxfam Emergency Aid Department

Oxfam has been working in emergency aid and disaster relief since its founding in 1942. In recent years the greatest part of Oxfam's funds have been spent on emergency work: in 1993-94 nearly 60 per cent of it was deployed in Africa. 'Conflict has been the main cause of our emergency and rehabilitation work. Most of Oxfam's emergency spending has gone to cope with the effects of human conflict, rather than natural disasters.' (Marcus Thompson, Emergencies Director). Oxfam is active in Africa, South and South East Asia, the Middle East, Latin America and the Caribbean, and in parts of Eastern Europe and the former Soviet Union. Much of Oxfam's emergency work is in assisting refugees, displaced people and host communities, providing relief supplies, particularly water and sanitation facilities, other public health measures, shelter materials, other subsistence needs (clothes, cooking pots, etc) and recently some psycho-social support to women in distress.

Marcus Thompson feels that there should be a 'more holistic approach to emergency relief, so that improvements in one sector are not undone by neglect of others'. Emergency preparedness should have a higher priority, and awareness of the needs for mitigation measures needs to be raised with major donors. The urgency of assisting the victims of high-profile and on-going conflicts is overshadowing the need to develop preparedness work for natural disasters.

Work and skills

The audit questionnaire identified two overlapping and complementary groupings: work content and skills. These have been combined into the following section.

Work content

Chart four below illustrates the proportion of activity stated in returned questionnaires in relation to the thirteen identified areas of 'work content'.

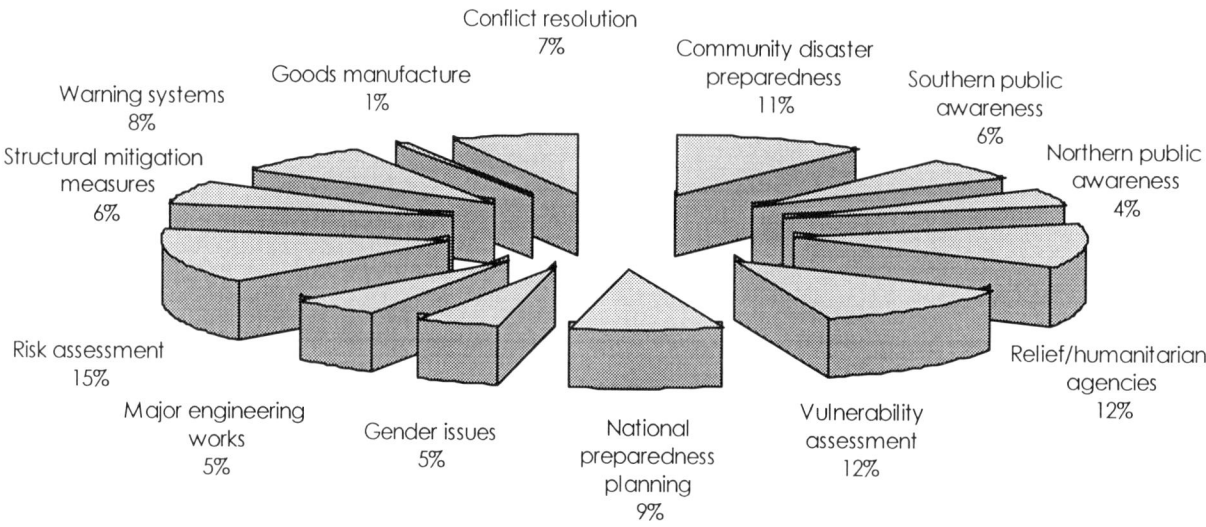

Chart Four   Percentage of work content according to organisation

The three highest areas of activity according to chart four are relief/humanitarian agencies (12 per cent), vulnerability assessment (12 per cent) and community level disaster preparedness (11 per cent) respectively. Two of these three relate to pre-disaster activity whilst relief clearly is post-disaster. It is surprising that relief did not rate higher, especially bearing in mind the large amount of relief activities, especially by NGOs, currently in progress.

In contrast to the high scoring areas of work content, low ratings were recorded for both southern and northern public awareness (6 and 4 per cent respectively), usually thought to be the domain of NGOs; gender issues and major engineering works (both 5 per cent) and, lowest of all, goods manufacture (1 per cent) registered as an activity only by government departments.

	Community disaster preparedness	Southern public awareness	Northern public awareness	Relief/humanitarian agencies	Vulnerability assessment	National preparedness planning	Gender issues	Major engineering works	Risk assessment	Structural mitigation measures	Warning systems	Goods manufacture	Conflict resolution
Charity/NGO	38	12	19	50	17	12	19	12	13	10	17	0	19
Private Company	31	23	23	15	38	38	15	8	54	23	15	0	15
Academic/research body	40	9	13	26	45	19	19	11	53	15	23	2	23
Consultancy	43	21	14	43	50	32	25	29	61	32	25	0	36
Individual Consultant	42	16	16	42	47	21	11	5	42	21	16	0	32
Government Department	14	29	0	43	29	43	0	29	57	14	57	14	14

Table Three Percentage of activities of organisations according to work content

Table three indicates the percentage of activities in different organisations according to work content. Striking statistics can immediately be seen. For instance, half of NGOs state activity in relief/humanitarian work (the highest percentage for any of the organisations, yet surprisingly low bearing in mind the growth in relief work for many of the larger agencies), and a high rating (38 per cent) is given to community disaster preparedness. However NGOs rate lowest in risk assessment at only 13 per cent. It is difficult to understand the low figure given to assessment; possibly regarding relief it may be argued that there is no time for such forward-planning activities.

Only 19 per cent of NGOs ranked northern public awareness as work content, lower than private companies at 23 per cent. However for the latter an interpretation may have been more to do with advertising of commercial concerns, whereas the former would regard advocacy and lobbying. Such a low NGO figure could be interpreted as disappointing when there are no other groups providing northern advocacy on fundamental development issues (which of course affects mitigation and preparedness).

NGOs also rated gender issues at 19 per cent, again surprisingly low when considering NGOs to be at the fore of advocacy. Interestingly consultancies placed gender issues highest at 25 per cent: possibly (if slightly cynically) a funder-led move, where funders require increasingly high profile adherence to gender issues. Again consultancies rated higher than all other organisational types for conflict resolution (possibly relating to long term post disaster activities).

A comparison of the four highest ranking areas of work content according to organisation is illustrated in Chart Five overleaf. From the chart it can be seen that the highest reported activity for consultancies was risk assessment followed by relief/humanitarian actions. In contrast returning NGOs recorded risk assessment as their *lowest* area of activity. Does this mean NGOs on the whole hire consultancies to carry out risk assessment?

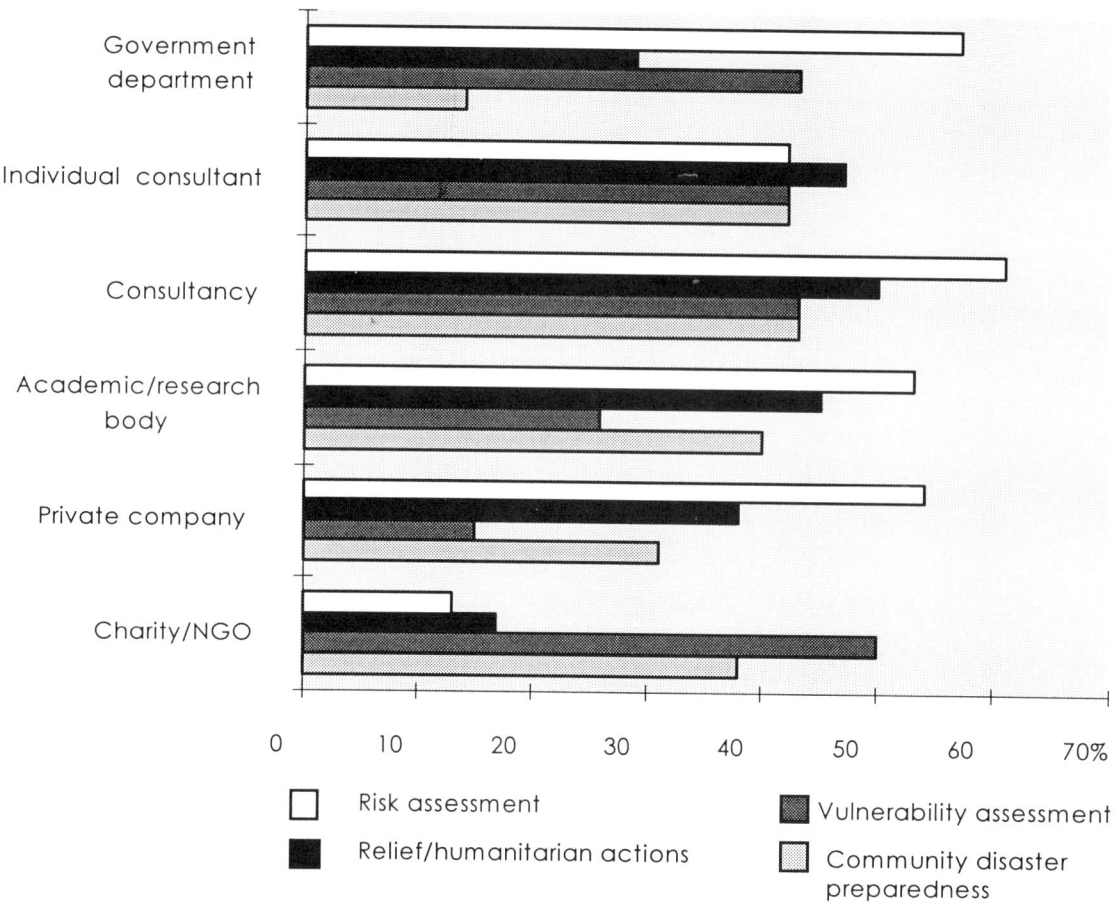

Chart Five Four key areas of work content according to organisational activity

Ove Arup
Engineers

Founded in 1946, Ove Arup is one of largest international firms of engineering consultancy. It currently employs a total staff of over 3 500 in 21 countries all over the world. They design buildings, bridges and other structures to withstand earthquakes, tropical cyclones, fire and other natural forces in many parts of the World.

Earthquake resistance is a major aspect of design in over 50 current projects, including many new buildings around the Pacific rim, retrofitting of buildings in California and Japan, provision of power stations in the Philippines, a centre of pilgrimage in Italy and a major transportation project in Bangkok. The company is also involved in seismic hazard and risk assessment for the Philippines, Hungary and the UK.

The partnership publishes the quarterly *Arup Journal*, and produced the recent textbook *Concrete Construction in Earthquake Regions* (Longman).

Skills

The skills section comprised twenty six categories, listed in table four on page 19. The skills listed were broad in coverage and definition. Chart Seven below illustrates the percentage of activity of organisations according to the five highest ranking skill areas. These are in order of ranking:

1. Training
2. Social science research
3. Technical research
4. Conflict prevention/tension reduction
5. Health/epidemiology/nutrition

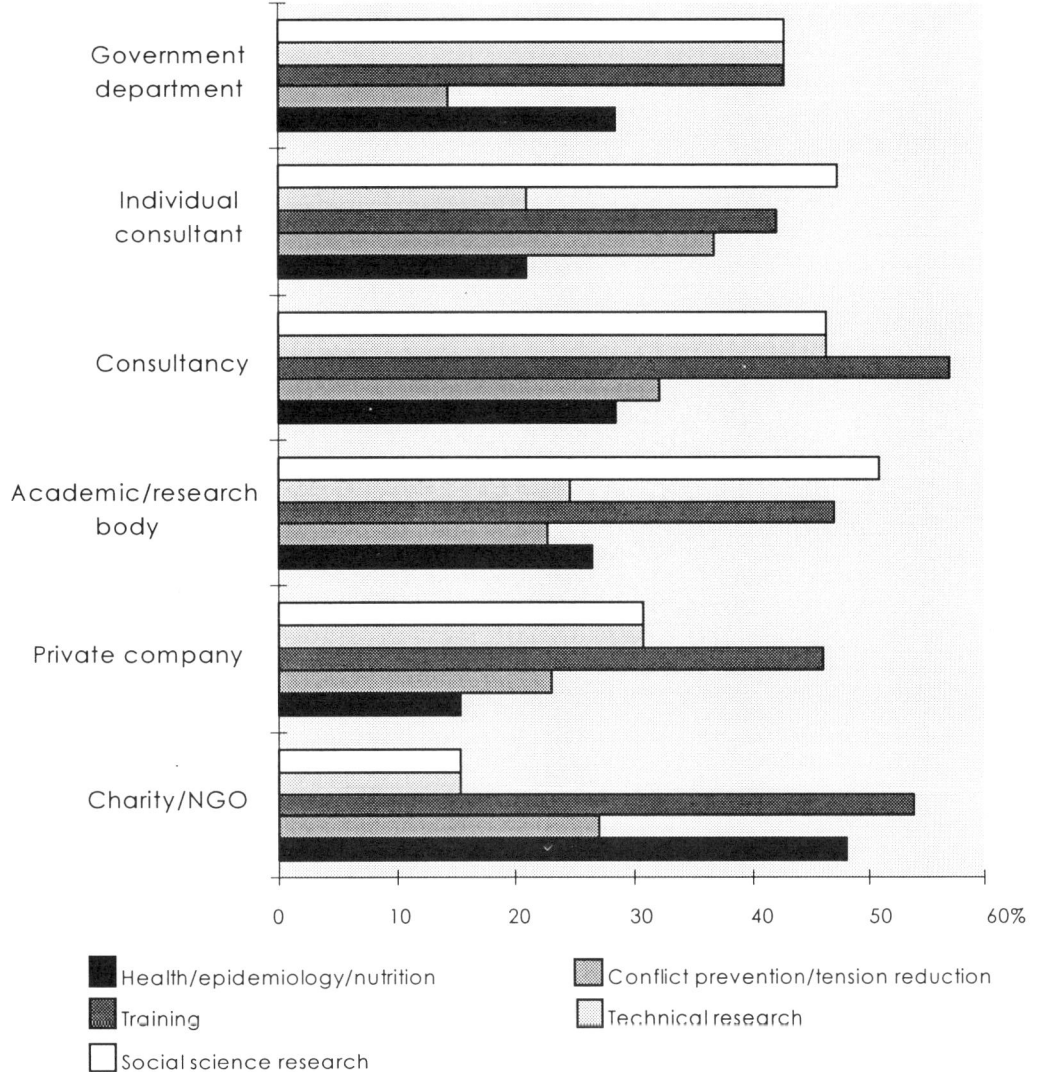

Chart Six Bar chart showing the five highest scoring skill areas according to organisational activity

All organisation types except for academic/research bodies and individual consultants ranked training as their highest category. With 'training' as such a broad category this is not surprising, also bearing in mind the current emphasis of developing local know how, often seen to be best achieved through training programmes. By referring to table four it can be seen that over half (57 per cent) of consultancies registered training as a skill area, whilst individual consultants were lower at 42 per cent, rating only social science research higher at 47 per cent. 54 per cent of NGOs registered training as a skill, whilst the second highest NGO category was health/epidemiology/nutrition at 48 per cent. These two categories were much higher than any other; no NGOs registered volcanology as a skill area.

What was surprising was the very low levels of NGO activity in research: 15 per cent for both technical and social science research. In contrast 46 per cent of consultancies claim to carry out research in these areas. A possible explanation of this could be that NGOs do not consider research a 'direct action' activity benefiting the poor, cannot afford it (for the smaller NGOs) or learn enough from southern partners to need not to attach the label research to their accumulating knowledge.

37 per cent was recorded for agriculture, possibly reflecting the traditional rural base of activity of most NGOs. This was the highest percentage for all organisational types: others were all below 30 per cent. Similarly low percentages, all 27 per cent and below were recorded for forestry. This is surprisingly, and possibly an area to be addressed amongst the DMP community, at a time when deforestation and food production are such critical issues.

Unsurprisingly private companies registered highest in building and architecture (38 per cent) and physical planning (46 per cent): activities often with high capital expenditure and technical expertise. Private companies also ranked highest in energy, the lowest scoring skill category, again a highly specialised activity. Government departments' key skill areas were in research (technical and social science) and communications. Hydrology and volcanology scored high (29 per cent). Almost one fifth (18 per cent) of consultancies stated activity in insurance/reinsurance; only 4 per cent of NGOs stated activity in this area.

Academic/research bodies ranked highest for social science research (51 per cent), followed by training (47 per cent), possibly indicating the growth of academic departments carrying out consultancy work at a time of diminishing university public funding. 26 per cent recorded activity in health/epidemiology/nutrition and AT/indigenous knowledge, followed by anthropology at 25 per cent. Academic departments tended to have a niche, eg London School of Hygiene and Tropical Medicine for Health and the Development Planning Unit at the University of London for physical planning.

Rob Stephenson
Independent Consultant

Rob has been in disaster related work since 1975 but now finds he has been increasingly shifting away from disaster and emergencies towards information science and library work. He finds that he is also shifting very rapidly towards more commercial clients outside the traditional emergencies network as he is assured of quicker responses in project development.

As an individual consultant Rob needs high-value, short-term work. He uses the gaps in-between projects either to expand on his previous project or to take stock of what he is doing. Proposals are developed 'on spec' with the understanding that sooner or later he may have a project. 'To be realistic this is a great deal of what goes on in terms of project proposals: repackaging.'

'The structure of this business has changed dramatically in the last year. There is a tendency for major donors to go through larger institutions and develop a cluster of consultants around them. The competition in these clusters is intense. The key thing now is to step out and look elsewhere for funding. There are many new consultants coming in from development, business and military backgrounds, looking towards relief work for employment. There seems to be very little collective attempt to develop strategic forecasting. A collective idea of where we might be going and what the world might look like in two or three years from now is largely missing. It is very difficult to lobby and influence the system when you are an individual.

'Overall the whole system looks like a Rubik's Cube and few people can see a good way in which to improve it.' Rob feels there is a need for more collective activity with an environment in which people from different groups feel more confident about exchanging information. Above all there is a need for access to project results and evaluations.

	Food Security	Engineering	Seismology	Building and architecture	Physical planning	Health/epidemiology/nutrition	Agriculture	Forestry	Hydrology	Geomorphology	Volcanology	Anthropology	Conflict prevention/tension reduc	Meteorology	Insurance/reinsurance	AT/indigenous knowledge	Energy	Remote sensing	Development economics	Transport	Training	Information management	Technical research	Social science research	Communications
Charity/NGO	35	19	8	21	12	48	37	27	13	6	0	13	27	8	4	21	8	2	15	21	54	25	15	15	25
Private Company	8	23	15	38	46	15	15	8	15	23	15	15	23	8	31	15	23	8	15	8	46	38	31	31	31
Academic/research body	28	15	8	21	13	26	21	8	11	13	13	25	23	8	8	26	4	13	23	6	47	15	25	51	9
Consultancy	36	39	14	32	32	29	29	4	18	14	7	25	32	7	18	29	21	4	25	18	57	36	46	46	25
Individual Consultant	37	5	5	16	26	21	21	5	11	21	5	26	37	11	11	26	0	0	11	5	42	11	21	47	21
Government Department	14	29	29	14	14	29	14	0	29	14	29	29	14	29	0	0	0	29	29	14	43	29	43	43	43

Table Four Percentage of skills according to organisation

The table indicates as a percentage the skills registered as present in specific organisations: for example, 35 per cent of all charities/NGOs returning the questionnaires indicated skills in food security.

Education and training courses

UK based academic courses

The audit aimed to gain an overview of development courses offering hazard/disaster elements as part or all of their course. The audit found a total of 31 UK-based courses or modules being taught with hazards offered as a topic within the course[9]. Of these there are 16 Master/Diploma courses, two institutions offering specialist PhD and MPhil programmes, whilst for undergraduates one new BSc course and four disaster-related modules within existing degree courses. A list of academic courses returned on the questionnaires is provided opposite.

Courses offered include agricultural management, nutrition and environmental assessment, engineering for structural damage, earthquakes, complex emergencies, built environment and health. The average length of postgraduate courses listed are one year, of which the diploma is taught and the masters dissertation is based on research.

A new undergraduate course offered by The Fire Services College at Coventry University, a BSc in International Disaster Engineering and Management, is due to begin in September 1995. The course 'will cater for those whose talents are directed towards solving social, technical and economic problems in disaster zones and for those aspiring to be future leaders in this field'. The three year course will include Engineering and technology, construction engineering, emergency management and courses in field skills.

The Centre for Urban and Regional Studies at the University of Birmingham is currently planning a course on Conflict and Complex Emergencies. An MA by research is also available for study in this area at the Post-War Reconstruction and Development Unit (PRDU) at York University. Remote sensing and Geographic Information Systems (GIS) appears to be a growth area in the geography discipline and it appears will occupy a significant proportion of education and research development. Institutions offering distance learning courses include The Open University and the International Extension College.

Oxford Brookes University
Complex emergencies and humanitarian aid

The Centre for Development and Emergency Planning (CENDEP) has developed an area of specialisation in Complex Emergencies as part of their Diploma/MSc Course in Development Practices. It is an example of the type of course being introduced in other academic institutions and it reflects the growing concern in this area.

CENDEP offers special study in this field consisting of four modules, each 10 weeks in duration. These include: an *Introduction to Complex Emergencies*, series of lectures and seminars exploring the main international policy issues involved in response to complex emergencies; *Methodologies in Emergency Practice*, a course of ten seminars led by visiting experts which look in depth at the main methodologies for effective UN and NGO practice in complex emergency, a terms project which is based on two weeks fieldwork; and *Environmental Hazards and Disasters*, a study of the response to natural and technological disasters in selected developing and developed countries. In addition students may take options in refugee studies at Oxford University.

---

[9] It is recognised however that many geography courses not included in the audit may contain hazard-related components.

UK based hazard related academic courses

The following list comprises UK based academic hazard related courses and modules as indicated by returned questionnaires:

Organisation	Course/Module Title	Duration
Centre for Arid Zone Studies, University of Wales	MSc Water Resources; MSc Rural Resource Management	1 year
Centre for Developing Areas Research (CEDAR)	Module of MA/MSc Ecology and Land Management in Latin America; Environmental Risk Assessment, module of undergraduate degree course; Tropical Ecological Systems, undergraduate degree module	1 year
Centre for Development and Emergency Planning (CENDEP), Oxford Brookes University	Diploma/MSc Development Practices. Areas of special study in environmental hazards and emergency practices	1 year
Centre for Urban & Regional Studies, University of Birmingham	Planning a course on conflict and complex emergencies	Planned
Climate Research Unit, University of East Anglia	MSc Climate Change	1 year
Department of Geography, University of Cambridge	MPhil in GIS and Remote Sensing	1 year
Department of Civil Engineering, Imperial College London	BSc/BEng Engineering Seismology/Soil Mechanics; BSc/BEng Earthquake Engineering and Structural Dynamics; BSc/BEng Structural Steel Design	1 year
Department of Geography, Chester College	BA Module in Natural Hazards and Environmental Management	15 weeks
	Research Studentships (MPhil/PhD)	2/3 years
	BA Module in Disaster and Development	15 weeks
Durham University Business School	MBA Module in Crisis Management	
Earthquake Engineering Research Centre, University of Bristol	MEng in Earthquake Engineering	1 year
ETC (UK)	BA Modules in Local Management of Natural Resources and Sustainability and Development Training; BA Module Natural Disasters	12 Weeks
Fire Services College, Coventry University	BSc (Honours), International Disaster Engineering and Management	3 years
Hazard and Risk Management Studies (HARMS), London School of Economics	MSc/BSc Options in Hazard and Risk Studies	1 year
London School of Hygiene & Tropical Medicine	MSc Nutrition and Needs Assessment in Emergencies; MSc, Conflict and Health module; MSc Human Nutrition	
Natural Resources Institute with University of Greenwich	Diploma/MSc Grain Storage Management	4-24 months
Post-War Reconstruction and Development Unit (PRDU), University of York	MA Post-Conflict Reconstruction	Planned for Oct 1996
School of Oriental and African Studies (SOAS)	MSc Remote Sensing and GIS	1 year
Sheffield University	MSc Human Nutrition	
Systems Group, The Open University	BA Module Complexity, Management and Change (part of Systems Failures Course)	1 year Part time

UK based training courses

The table on the following page summarises UK-based short courses and training seminars as entered on the returned questionnaires, resulting from the 30 per cent of questionnaire respondents who stated that they organise UK-based training courses. Some courses are one-off whilst others are held regularly. Length ranges from one day to several weeks. Most of the courses do not offer a qualification but do present an attendance certificate.

The table illustrates the breadth of courses offered, including hazards, risk and safety, water supply techniques, refugee issues and codes of practice for wind loading. One of the longest and largest courses is the annual four-week Summer Disaster Management Training Course offered by the Disaster Preparedness Centre at the University of Cranfield.

Overseas training capacity

Overseas training is clearly a large area of activity for all organisations, as discussed in the *Work Content and Skills* section. In answer to the direct question on the questionnaire of whether respondees were involved in training, 43 per cent of respondents considered themselves training providers either in the UK and overseas, whilst 44 per cent of respondents listed training as a skill in their organisation. These results confirm that the training capacity in the UK maintains a high priority. This could be related to recent funding trends and the belief in local capacity building as a key activity.

Only a few organisations specifically specialise in training. One recently-formed organisation, the Oxford based International NGO Research and Training Centre (INTRAC), provides training both in the UK and overseas. From the current activities listed from the respondents it is clear that there are numerous projects which are training related or have it as a component part of their project in the form of seminars and workshops (the *Current Activities* section provides an indication of this).

The London School of Hygiene and Tropical Medicine
Research and Education in Conflict, Nutrition and Health

The London School of Hygiene and Tropical Medicine's (LSHTM) work in the field of emergencies began in the early 1980s with the establishment of the Refugee Unit. Today it works widely in research, consultancy and academic and training courses. Current research and consultancy projects include post-conflict health policy issues in Africa (Ethiopia, Uganda, Eritrea), Palestinian Territories and El Salvador; political violence and the maintenance of services during political unrest; transport and traffic disasters, for example bus crash disasters in Zimbabwe; and community-level nutritional needs assessments.

A key strength of the LSHTM's involvement in conflict and health is that it has staff members who come from a wide variety of different disciplinary backgrounds: public health, nutrition, epidemiology, water and sanitation, health economics, health services research, social policy and policy analysis.

LSHTM also runs a variety of in-country training courses including a two-year programme of short field courses for NGOs in nutritional needs evaluation, and the rehabilitation of malnourished children (for CIDA). In-house education courses include five-week modules in Conflict and Health, and Nutrition and Needs Assessment in Emergencies. Both modules form part of an MSc programme. Last year LSHTM inaugurated a fortnightly open seminar on current issues regarding conflict and health. Past talks have included NGO roles in refugee camps following the Rwandan Crisis, land mines and population displacement in Mozambique.

LSHTM has an accessible library with a large body of information on conflict, health, famine and drought. It is currently developing a two-year project for the placing of this information on CD ROM.

UK based hazard related training courses

The following list comprises UK based training courses as indicated by returned questionnaires:

Organisation	Course	Duration
AIDS, Care and Training (ACET)	AIDS Awareness and Training	variable
British Association for Immediate Care	Medical Management of Major Disasters; Basic Training Course for Overseas Delegates; First Aid Training	2 day, 3 and 7 day courses
Cargil Attwood Consultants	A wide range of communication skills courses including Project Management and Teamwork	variable
Catastrophe Reinsurance	Various conferences and seminars	variable
Centre for International Health	Working in Emergency and Disaster Relief	1 week
Climatic Research Unit	General climate change analysis/modeling	variable
Cranfield Disaster Preparedness Centre (CDPC)	Disaster Management: introduction and focus on Slow & Rapid onset & Technological Hazards and training.	6 Weeks
Cranfield School of Management	Strategies for Change: Managing NGOs	3 Sept - 6 Oct 95
Earth Resources Centre	Hazardous gases	
ETC (UK)	Sustainability and Development Management	14 days
Geology Department, University of Bristol	Geological Fluid Mechanics	4 weeks
Hadley Centre Meteorological Office	Meteorological Training Courses	variable
Institute of Development Studies (IDS), University of Sussex	Food Security Short Courses	variable
International NGO Training and Research Centre (INTRAC)	Organizational Assessment for NGO Capacity Building; Financial Management For Non-Financial Managers; Participative Techniques for Monitoring and Evaluation	Jun-Oct 95
Liverpool School of Tropical Medicine	Short Courses	variable
Mott MacDonald Group	Hazard, Risk and Safety	1 day
Oxford Centre for Disaster Studies (OCDS)	Disaster Management Training of Trainers	August/ September
Overseas Development Group	Visiting Professional programme Short Courses	variable
Post War Reconstruction & Development Unit (PRDU), University of York	Practical Lessons on International Disasters Interventions	1 week
RED R Registered Engineers for Disaster Relief	Water Supply and Survey - inc simple intake	27-30 Jul 95
	Field Construction - Temporary shelters, etc	6-8 Oct 95
	Needs Assessment Needs after disasters	27-29 Oct 95
	Refugees, Agencies and Relief Workers	12-17 Nov 95
Responding To Conflict	Group Development and Problem solving	
	Working with Conflict (5 Modules)	11 weeks
	Facing Violence: Strategies for Social Change Towards Just Outcomes	
Wind Engineering Society	Use of code of practice for wind loading	1 day
World Association for Disaster and Emergency Medicine	Medical Management of Major Incidents (at The Royal Postgraduate Medical School, Hammersmith Hospital)	3 days

Networks

Formal networks, professional or electronic, do not play an important role in the life of the British disaster community, or so the replies to the questionnaires indicate. Although over a hundred respondents were involved in a network, a similar number of different networks was listed, and few organisations or individuals were linked to more than one.[10]

Questionnaire results

However, the returns should be handled cautiously. Those who replied to the question about professional networks interpreted it in many various ways.

When filling in the questionnaire form, respondents took a very broad definition of the term 'professional network'. This was understandable given that the form itself did not define the term any further and that the expression 'network' has always been used fairly loosely to cover a number of activities from sharing information to collaborative ventures and lobbying. The result, then, was a very diverse list of networks based in the UK and overseas (a full list of their names is attached in Appendix Two).

The British networks or associations listed can be grouped under the following main headings:

- Those focusing on disaster issues or themes such as the ODI's Relief and Rehabilitation Network, the Institution of Civil Defence and Disaster Studies or the UK Working Group on Landmines;

- Professional/technical associations whose members' interests may encompass disaster work. This category includes such organisations as the Royal Meteorological Society, Institution of Civil Engineers, British Nutrition Society and Association of Geoscientists in Development;

- Geographical interest groups (sometimes with a disaster angle), for instance, the European Network of Bangladesh Studies at Bath University and the British Agencies Afghanistan Group;

- Networks of agencies working in development, including British Overseas ngos for Development (BOND) and the Development Studies Association, as well those focusing on particular development issues such as structural adjustment or pastoralism.

Because the question was open, respondents were left to make their own decisions about which networks to mention. Some thought they should refer to disaster networks only or at least to networks whose areas of interest covered disasters specifically, whereas other lists covered a much larger field.

While the returns certainly underestimate the extent to which organisations and individuals are involved in networks of all kinds they may be more accurate in pointing to the degree of participation in disaster-related networks. Even here, though, the replies referred to formal networks. Informal networking arrangements (and there are certainly some, perhaps even many) remain invisible.

The replies also included some electronic networks. Although electronic systems were covered separately in the questionnaire, the apparent inconsistency of the responses here probably reflects the impossibility of drawing a neat distinction between professional and electronic networks.

Meaningful statistical analysis of the results is clearly not possible. Therefore, the discussion here concentrates on a few seemingly significant facts and key issues, outlining some significant disaster networks in operation and making suggestions about needs and opportunities in this field.

---

[10] The questionnaire format appeared to leave room for three entries only but this does not appear to have been a constraint.

Disaster networks

From the replies to the survey, three disaster networks appeared to be important in terms of the number of respondents who were members. Outlines of these are given here. As noted, the replies may undervalue the importance of other networks.[11]

UK network on conflict, development and peace (CODEP)

CODEP was established in 1991 in response to the growing need for NGOs working in conflict to share information and identify areas of particular concern. Its main focus is on analysing conflict, development and peace as they relate to NGO operations in the field, and hence on determining effective strategies for working in conflict and contributing to peace. Dissemination of these findings, and advocacy among the development community, are essential adjuncts to this work.

CODEP currently focuses on Africa but does not rule out a wider sphere of interest in the longer term - and certainly welcomes information and ideas from those dealing with conflict elsewhere.

It has a mailing list of nearly 40 'members', although the active membership attending meetings is smaller. Most are NGOs operating in the field or campaigning on issues but academic institutions, researchers and other specialists can also take part. CODEP does not produce a formal newsletter or publications but it did help organise a major workshop on conflict and development held in Birmingham in November 1994.

Taking a sensibly pragmatic approach it has chosen to evolve gradually in response to members' needs, beginning with sharing information and moving on when appropriate to synchronising work more efficiently and then, perhaps, more formal co-operation. Recently it has discussed the desirability of setting up a secretariat.

For further details contact:

>Judy El-Bushra
>ACORD
>Francis House (3rd Floor) Francis Street London SW1P 1DQ
>Tel: 0171 828 7611/7612/6544/65 Fax: 0171 976 6113

Society for Earthquake and Civil Engineering Dynamics (SECED)
Professional network

SECED is the British branch of the European Association of Earthquake Engineering. In addition to the Society's interest in earthquake engineering, its activities embrace the wider sub-discipline of civil engineering dynamics. The society plays an active role in building code issues, research and education in the field. To fulfil its diverse role SECED has four technical reporting groups covering civil engineering dynamics, engineering seismology, earthquake engineering and soil dynamics. Each technical reporting group is responsible for monitoring and reporting progress in the relevant field and making proposals for action by the society.

The activities undertaken by SECED include the SECED Newsletter, monthly technical meetings, biennial Mallet-Milne Lecture, triennial technical conference, occasional workshops and seminars. SECED's most notable assets comprises an extensive directory of practitioners.

The SECED membership comprises professionals from industry and academia. It is associated, in the UK, with the Institution of Civil Engineers, Mechanical and Structural Engineers, the Geological Society and with the Wind Engineering Society.

---

[11] The UK Working Group on Landmines, for instance, has some 30 members, including NGOs, pressure groups, funders and others, several of whom replied to the questionnaire but only one of whom noted its membership.

Refugee participation network (RPN)

Part of the Refugee Studies Programme at the University of Oxford, the RPN is a forum for researchers, refugees and people who work with refugees to exchange experience, information and ideas, and to debate issues. It aims to provide practical information to those working in the field, extend their understanding of refugees' needs and aspirations, and encourage refugee participation in the design and implementation of projects that are meant to benefit them. By disseminating research findings in a way that makes them relevant to practitioners, research can serve to influence policy and improve practice.

The main medium for this work is the thrice-yearly publication *RPN*, which is sent to 2100 members in 110 countries. A directory of members is also available. Membership subscriptions are encouraged.

The Refugee Studies Programme's documentation centre has over 20 000 documents including both published and 'grey' literature, and may soon be accessible through the Internet.

Information is available from:

> Marion Couldrey,
> Co-ordinator
> Refugee Participation Network
> Refugee Studies Programme
> Queen Elizabeth House, 2 St Giles, Oxford, OX1 3LA, UK.
> Tel: 01865-270722   Fax: 01865-270721 e-mail: rsp@qeh.ox.ac.uk

Relief and rehabilitation network (RRN)

The RRN was set up in 1993 with funding from EuronAid and the European Commission. Recognising that NGOs are not able to learn much from others' experiences, its main objective is to improve the flows of information between NGOs involved in relief and rehabilitation, and between NGOs and researchers.

Its membership is composed mainly of NGOs working in the field. Individual NGOs nominate their personnel, including those based in head offices, who are directly involved in planning and implementing relief operations. Individuals and personnel of government and UN agencies are also able to become members, although at a higher fee.[12] By March 1994 the RRN had 132 individual members.

Members receive regular newsletters and network papers, and also commissioned 'state of the art' reviews by specialists of different areas within the relief and rehabilitation field (the first two were on water and sanitation, and emergency supplementary feeding programmes). They can also obtain advice on technical and operational problems from the ODI or other members. A register of members detailing their disciplines, areas of experience and interests has been produced.

For details, contact:

> Véronique Goëssant
> Relief and Rehabilitation Network, Overseas Development Institute
> Regent's College, Regent's Park, Inner Circle, London, NW1 4NS, UK.
> Tel: 0171 487 7413   Fax: 0171 487 7590
> email/Internet: odi@gn.apc.org (mark for attention RRN in subject line)

UK IDNDR committee and working groups

Only three respondents cited these as networks, yet they clearly perform a networking role in transmitting information and perhaps in stimulating collaborative work. Further enquiries among members of these groups might throw more light on this.

---

[12] The fee structure is graded depending on the type of organization and number of staff within it who are members.

Professional/technical associations

The responses threw up a welter of professional and technical associations and societies, some based around broad themes while others were quite specific in their interests. They covered such subjects as dams, hydrology, nutrition, geography, geology and psychology; and they included professions such as civil and consulting engineers, architects and management consultants. With such diversity it was impossible to group them into a clearly definable category or list.

Only three of these 'networks' had more than one member among those who returned the questionnaires: the Association of Geoscientists in Development (AGID), Institution of Civil Engineers and Royal Institute of British Architects. Here too, though, the returns may be misleading. It is probable, for instance, that some individuals working within organisations replied only on behalf of their employers and did not record their personal membership of professional institutions.

Even if we cannot gauge the extent of their importance, it is safe to assume that associations of this kind can play a significant role in transmitting ideas and information beyond the confines of the disaster community itself and out to those whose work may from time to time involve preparing against disasters or dealing with their results. For this reason it might one day be useful to carry out a more detailed piece of research exploring the full range of professional and technical societies and their links with other networks and journals.

Geographical networks

Six geographical networks appeared from the survey although two are outside the UK. The four based in Britain are the British Agencies Afghanistan Group (BAAG), European Network of Bangladesh Studies, Inter-NGO Committee on Somalia (INCS-Forum) and Sudan Lobby Group. The two based overseas are the Agency Co-ordinating Body for Afghan Relief (ACBAR) in Pakistan/Afghanistan, and the Bangladesh Disaster Forum in Dhaka.

The list could certainly be expanded from further research. Networking and co-ordinating associations are often created in response to emergencies, especially conflicts, affecting specific areas. The Gulf Information Project is one: it arose in 1991 at the initiative of the Refugee Council as a result of the war and large numbers of refugees from Iraq. In 1994 agencies came together very quickly to share scarce operational information and co-ordinate actions to address the crisis in Rwanda.

Development networks

There are a number of UK networks for those working in development. Again, there are three particularly important to the UK disaster community, to judge from the returns.

British overseas NGOs for development (BOND)

BOND was formed in June 1993. It is a broadly based network with about 100 paying members, all UK-based voluntary organisations, who are working in 160 countries world-wide. The 12-person executive committee is balanced so that small and medium-sized NGOs are adequately represented. Associate membership is open to other types of organisation interested in promotion of or research into issues concerned with overseas development and development education. Its growth in membership and range of activities is eloquent demonstration of its perceived value to the NGO community.

Its aim is to enhance the effectiveness of development assistance from the UK by sharing experience and ideas, among its members and with the British Government. It has produced a directory of members and maintains a database of their interests. A quarterly newsletter is circulated covering all kinds of development themes and events including humanitarian aid and shelter.

BOND has set up working groups on key issues (currently project evaluation, ODA funding mechanisms, environmental impact, NGO management and collaboration with southern NGOs) and also runs workshops on these and other topics (among the workshops at the May 1994 general assembly was one on 'emergency assistance and support for livelihoods'). The programme is meant to expand in response to members' interests. BOND has also enabled the NGO community to lobby on issues of common interest.

For details contact:

> British Overseas NGOs for Development
> Regent's Wharf, 8 All Saints Street, London, N 9RL, UK.
> Tel: 0171 713 6161   Fax: 0171 713 6300

Development studies association (DSA)

The DSA was set up in 1978. It exists to advance knowledge of development issues through meetings, conferences and disseminating information about research results, operation and practice, course curricula and training schemes.

Its members include academics and development practitioners of different kinds (individuals and organisations). A subscription fee is payable, the amount paid depending on the type of member. Much of the DSA's work is founded upon several study groups of which two are relevant to disasters: international hazards; and development, disarmament and security (incorporating the former war and development group).

For the International Hazards Study Group contact:

> Ewan Anderson
> Department of Geography, The University, Durham, DH1 3LE, UK.
> Tel: 0191 374-2448

For the Development, Disarmament and Security Study Group, contact:

> Geoff Tansey
> 4 St John's Close, Hebden Bridge, West Yorkshire, HX7 8DP, UK.
> Tel: 01422 842752
> or,
> Paul Rodgers
> Department of Peace Studies
> University of Bradford, Bradford, BD7 1DP, UK.
> Tel: 01274 733466

DSA runs a conference of two to four days each autumn around a particular theme: in 1995 this is 'Denying famine a future: new approaches to food security, development and aid'; among earlier themes was 'Conflict and change in the 1990s'. The DSA also publishes a newsletter which reports on discussions at the study groups, is a platform for debate on issues, and contains the usual details of events and publications. DSA is also responsible for the *Journal of International Development.*

For information, contact the Membership Secretary:

Dr Michael Hubbard
Development Administration Group, ILGS, University of Birmingham, Birmingham, B15 2TT, UK.
Tel: 0121 414- 976; e-mail: dsa@bham.ac.uk

EC-NGO network

Each member state of the European Union (EU) has its own platform of development NGOs interested in closer links with the EU's institutions. The national platforms are represented through the EC-NGO Liaison Committee based in Brussels (see *Funding*).

The UK platform holds three general assemblies a year. Many if not most British development NGOs are now members. There are working groups on development policy, development finance and development education. There is also an ad hoc group on VOICE; this brings together NGOs who have significant dealings with the European Community Humanitarian Office, VOICE and the ODA. The network has lobbied repeatedly on food aid issues over the years.

For further information contact:

> Mike Aaronson
> UK representative, EC-NGO Network
> Save the Children Fund, Mary Datchelor House, 17 Grove Lane, London, SE5 8RD, UK.
> Tel: 0171 703 15400

Conclusions: problems and potentials

What can we deduce from the apparently low level of formal networking among the UK disaster community? That it is not necessary or not thought to be necessary?

Interpretation

One explanation may be that because informal, less visible networking is widespread there is no need for networking institutions. Another is that the community is more interested in smaller, specialist groups than in multidisciplinary networks; or that it is happy to see disaster issues included within other associations' spheres of interest.

It is more likely that the root of problem lies in the diverse membership of the disaster 'community' itself. This is not really a community at all but a large collection of individuals and institutions specialising in particular aspects of disasters. The diversity of hazards is matched by the range of specialist skills needed for mitigation and relief work. It is not surprising that successful networks have focused on single themes such as refugees, conflict or particular complex emergencies. Nor should we be surprised that attempts to create more wide-ranging discussions (such as the International Hazards Panel) have had a chequered history.[13]

The need for networks

The rational argument in favour of networking remains unanswerable. Two main reasons are:

- At one level networking improves access to and exchange of information and expertise;

- Beyond that, it can help network members to maximise their impact through partnerships and greater co-operation; what is commonly called 'synergy'.

To this we can add a third overriding aim:

- To support vulnerable communities by responding to their needs. With the growing frequency, impact and complexity of disasters internationally, especially in the South, there is an ever more pressing need to make interventions as informed and effective as possible.

Everywhere in the world networks are establishing themselves and flourishing, with encouragement from governments, international agencies and donors - especially for multi-disciplinary activities. Even in the UK, where there is not a strong culture of sharing and co-operation across the disaster community, the networking urge is growing. Two meetings organised over the past year by the IDNDR Working Group on Application and Implementation, both designed to encourage information sharing and debate, drew good audiences from many parts of the community.

---

[13] The International Hazards Panel was set up by Intermediate Technology and others in 1980 as a forum to bring together people with separate interests and engaged in varied disaster-related activities. Diminishing interest and commitment from the community finally forced it to wind up in 1991, when it was absorbed into the DSA's Hazards Study Group.

The way forward?

The main question that should concern us is not the need for networking itself but the most appropriate means of meeting the different needs that networks fulfil. Many options are open. Here we suggest a three-phased programme (the phases may overlap). This is emphatically pragmatic. It allows the network to develop in response to demand and also represents a cautious approach to the question of financing. The phases would certainly take years to complete.

Phase 1: sharing information

This should start now. For exchanging information an electronic networking system may be sufficient at this stage. This would consist of a bulletin board open to all or creation of a more restricted user group. The power of such a system would be greatly increased if documentation centres now extant were on line.

The kind of information disseminated would include news of publications, papers and other materials; meetings, workshops and conferences; skills and capacities available; current and planned projects; partners or expertise needed; job vacancies; sources of funds, recent grants made, funders' policies and deadlines; and new developments within major agencies. To follow up on a given item users would contact the organisations or individuals concerned.

Electronic systems are the most flexible and adaptable method of sharing information, and responding to interest and demand. It is also the best, possibly only, way of embracing the multidisciplinary nature of disaster work. This arrangement might have to be supplemented by distribution of printed information in the short term until use of the Internet becomes more widespread.

At this stage the network could become parasitical and try to attach itself to somebody else's newsletter or to an existing network or information service. In any case it should make links to existing facilities from the outset.

Phase 2: meetings and partnerships

To debate and discuss ideas it will be necessary to arrange meetings, workshops and the like. This requires some kind of co-ordinating mechanism but may only need to be fairly limited.

It is impossible to envisage particular activities to stimulate partnership at this stage, since it is assumed that partnership will grow out of information sharing and a change in culture towards co-operation. This could include joint projects in the field or concerted lobbying efforts. Should the network (or indeed networks) reach this stage a more substantial institutional home will be essential.

Phase 3: institutional

From the beginning the network(s) will need some kind of focal point; just where does not matter much as long as it has the requisite infrastructure and communications links. As it develops a more substantial programme of activities and even a more formal, institutional identity it will need to find a home.

The difficult question, faced by every network at some point, is when to become an institution of this kind. At what point does a secretariat become necessary or desirable? In every case the functions of a secretariat (liaison with members, organising meetings, distributing information) have to be taken on by somebody. The debate is around the status of that support: formal or informal, temporary or semi-permanent?

NGOs and other institutions can provide homes for networks that begin as ad hoc, limited or temporary arrangements but then need a secretariat or other formal structure.[14] In most cases, embryonic networks need some support from one agency or another in order to establish themselves: this usually takes the form of staff time, meeting rooms and communications. The most suitable home for a multidisciplinary disaster network in the UK would be an organisation that is already sufficiently large, well established and equipped to take on organisational, information storage and information-sharing functions.

---

[14] The Refugee Council hosts the British Afghan Agencies Group, for instance, while ACORD has provided CODEP's administrative support.

Beyond the IDNDR

Networks are born to meet needs and evolve or die according to the extent they can adjust to changing circumstances. Such are the needs at present that we can expect there to be a multi-disciplinary network for a long time. Its sustainability can be guaranteed in part by the strength and flexibility of its organisational arrangements but the real key is the involvement of network members.

The IDNDR Committee and its working groups could play a valuable role in maintaining the networking momentum during the rest of the decade. Perhaps the UK IDNDR structures are in themselves an embryonic network, in which case it may be time to consider how they will develop during the rest of the IDNDR and what will succeed them.

Funding

One of the main tasks of the audit was to assess current sources of funding in the UK. It was not easy to identify funders. Few have a stated commitment to mitigation and preparedness. Recipients of funding were never likely to help the survey much: organisations already receiving money for disaster mitigation tend to be secretive about their sources for fear of competition.

Nonetheless, a separate questionnaire was sent to 160 possible donors, mostly charitable trusts and foundations but also some grant-making NGOs. Only nine replies were received[15] although a few sources were identified from responses to the main questionnaire. Many more who may be supporting mitigation and preparedness to some extent through assistance to long-term development projects perhaps automatically equate 'disaster' with 'relief' only, an attitude encountered widely during the survey of active agencies.

The questionnaires clearly do not allow any general conclusions to be drawn. However, additional desk research and discussions have enabled us to compile not a complete picture but certainly a preliminary sketch of the UK funding 'scene'.

The following sections outline some of the main sources of funds, with comments and analysis, indicating where further information can be obtained. Some general comments and recommendations make up the conclusion.

The Overseas Development Administration (ODA)

The ODA is by far the largest source of funds in the UK for all activities related to disasters. A significant feature of the British Government's aid programme, and of official development assistance generally, is the increasing proportion of humanitarian aid in the total. The same trend is visible within the ODA's country programmes.

The ODA's funding for relief activities has risen sharply in recent years from 2 per cent of its bilateral budget in 1982/3 to over 10 per cent in 1992/3, in response firstly to natural disasters (especially in Africa) and more recently to the growth of complex emergencies.

In the past, relief and development were dealt with by separate parts of the ODA and contacts between departments were infrequent. This is now changing with mounting recognition that the two types of activity are interdependent. In 1993/4 the ODA gave over £260 million to relief, refugee, food aid and preparedness activities: £179 million was bilateral aid, £68 million was spent by European Commission programmes, and nearly £13 million went to the United Nations.

Economic and social division (ESD)

This sponsors a considerable amount of economic and social research, its focus including:

- Funding research projects and dissemination of results through the Economic and Social Committee on Overseas Research (ESCOR). ESCOR has been an important source of funds for institutions engaged in research on socio-economic aspects of disasters (for instance, famine coping strategies);

- Funding research, information resources and related activities at the Institute of Development Studies (this has included work on food aid and food security);

- Supporting the information and dissemination activities of the Overseas Development Institute, including some of its work on disasters.

(continued on page 33)

---

[15] Three were from trusts that were not, in fact, supporting work on disasters.

Emergency Aid Department (EMAD)

Increased expenditure has been reflected in an expansion in staffing within the Emergency Aid Department, the main ODA section working on disasters. Before 1991 EMAD was exclusively a funding body channelling money for emergency and relief operations through multilateral agencies such as the United Nations and European Commission, and bilaterally through NGOs and recipient governments. In response to the growing number of disasters, the growing insecurity of relief operations and the Kurdish refugee crisis, the Disaster Relief Initiative was established in 1991 which allows EMAD to manage its own operations in the field. The Department also gives support to agencies helping refugees.

To help NGOs applying for support for humanitarian assistance projects, EMAD has published a set of guidelines on project proposals and reports. This also contains the names and phone numbers of key contacts. For details and general enquiries contact EMAD (ODA, 94 Victoria Street, London SW1E 5JL, UK; Tel: 0171 917 0273/0348; Fax 0171 917 0502).

Disaster Mitigation and Preparedness

EMAD has a separate section dealing with disaster preparedness. It has £2 million at its disposal to fund projects in the 1995/6 financial year (approximately 1 per cent of the ODA's total disaster budget). The fund is open to organisations and individuals of all kinds including international and government agencies, national and international NGOs, and consultants.

There are few limits on the kind of activity eligible for support under the scheme, provided that it has a definite output. Innovative projects and approaches are encouraged. Pure research is excluded since other ODA funds are available for this (see below). In some cases, where the ODA's geographical desks also have an interest in a particular area of work, inter-departmental funding is possible.

In 1994/5 over 40 projects were funded from the mitigation and preparedness budget. These include:

- Production of a field manual for emergency relief workers (by the Save the Children Fund) developing the use of radio in preparing against disasters in West Africa (Cranfield Disaster Preparedness Centre);
- Enhancing disaster mitigation networks in Latin America and South Asia (Intermediate Technology);
- Mapping volcanic hazards in Chile (British Geological Survey and Bristol University);
- Disaster management training in Turkey (Oxford Centre for Disaster Studies);
- Compilation by UNDHA of an emergency stockpile register;
- The development of a public information programme for IDNDR;
- Cyclone and seismic hazard mitigation training for government officials in Vietnam;
- Hazard assessment of possible flooding from dangerous mountain lakes in Nepal;
- The development of a legislative programme for disaster preparedness planning in The Caribbean for the Caribbean Disaster Emergency Response Agency.

There are no formal guidelines for applicants yet but these are being considered .

(Continued from page 32)

Sub-Saharan Africa and South Asia are favoured locations for studies. Research projects must produce results that merit wide dissemination, have direct policy implications and will be of practical relevance.

Types of economic and social research considered important by ESD are:

- Analytical research into causal relations in an area of current development policy;

- Illumination of emerging policy issues;

- Evaluation of a particular practice or operation;

- Testing and developing new theory;

- Developing new research methods.

Priority interests are set out in ESD/ESCOR's three-year research strategies. A new strategy for 1995-8 was being prepared as this audit was being written and should be available by the time this appears in print. The likely interests were:

- Enhancing productive capacity through economic liberalisation and private sector development;
- Promoting policies and practices to reduce poverty;
- Responding to the challenges of human development;
- Understanding environmental change;
- Interactions of the state and society.

The ODA intends to commission up to 10 specific research programmes addressing its main interests, through competitive tendering. Themes will be advertised to find out who has an interest, the capabilities of applicant institutions will be assessed, and on this basis consortia of institutions will selected.

However, ESD/ESCOR has always supported a substantial amount of research outside its areas of interest (an 'open programme'), and will continue to do so under the new strategy. It has funded work into the politics of humanitarian interventions in the past.

Any UK-based researcher or institution may apply to ESCOR. Grants can be made to researchers in the South only in collaboration with a British institution. Guidelines and details of how to apply are obtainable from:

> Room V 736
> Economic and Social Division, ODA
> Tel: 0171 917 0636 (for general enquiries); Fax: 0171-917-0734

Regional/bilateral funds

The ODA's geographic (country) desks in the UK and Development Divisions in the South may, on occasion, make funds available for disaster mitigation work of different kinds, depending on the ODA's development priorities in the particular region or country.

Long-term emergencies and refugee projects are dealt with jointly by EMAD and the relevant geographical department.

Telephone numbers for general enquiries to the geographical departments are as follows:

Africa

- Central and Southern: 0171 917 0435
- Eastern: 0171 917 0434
- West and North: 0171 917 0467

Asia/Pacific

- Eastern Asia: 0171 917 0308
- South Asia: 0171 917 0358
- Western Asia: 0171 917 0343

Latin America/Caribbean/Atlantic

- General: 0171 917 0248

Health and population division

The Health and Population Division has provided occasional funds for those involved in work on nutritional aspects of relief and food aid: these include the Nutrition Unit at the London School of Hygiene and Tropical Medicine. It also supports several programmes of medical and epidemiological research in UK institutions.

The phone number for enquiries is 0171 917 0107

Engineering division

Scientific and technical research and development is the responsibility of the Engineering Division, which has given generous support to disaster research in the past.

Disaster-related grants in recent years have included work on earthquake-resistant housing design (Cambridge Architectural Research), analysing flood data for use in designing control projects (Institute of Hydrology), and the development of megacities and their vulnerability to disasters (Institution of Civil Engineers).

As part of the Division's work in architectural and physical planning, funds were given to the recent 'Building for Safety Initiative' which aims 'to bring the knowledge of how to build safely to those who need that knowledge most'. This included a series of four books published by IT Publications, each on a different aspect of the subject.[16]

The current strategy for Technology Development and Research (TDR) runs for three years. The Division appears to have around £8 million to spend each year on TDR projects, grants ranging from £50 000 to £300 000. This scheme is open to all UK organisations, voluntary and commercial.
There is an annual applications cycle: notices inviting proposals are published in June, applications submitted in September, and selection in January for funding from April. There were 198 TDR applications in 1994 for work in 1995 (the first year of the current scheme) and the ODA described the competition as 'extremely fierce'.

In its 1995 grants the programme focused on five main areas, of which three appear to offer some funding possibilities for those working in disasters:

- Water and sanitation (including management systems for flood control and prevention of drought, and making water available for sustainable food production);

- Geoscience (including identifying and mitigating geochemical toxic hazards, and improving geotechnical hazard avoidance strategies in national planning);

- Urbanization (including the provision of housing and infrastructure).

Energy efficiency and transport are the other main themes. The detailed TDR theme objectives are reviewed annually and may be altered. For this reason, and because there are additional definitions and limitations within these areas, potential applicants should discuss their ideas with the ODA in advance.

Information on the TDR strategy and funding can be obtained by contacting the Engineering Division, Room V362 at the ODA (Tel: 0171 0917 0484; Fax 0171 917 0072).

ODA evaluation department

The Evaluation Department has, from time to time, commissioned studies of the ODA's response to particular emergencies. The enquiry point for the department can be reached on 0171 917 0545.

---

[16] A. Clayton and I. Davis, *Building for Safety Compendium*; A. Coburn *et al*, *Building Principles for Safety*; E. Dudley and A. Haaland, *Communicating Building for Safety*; Y. Aysan *et al*, *Developing Building For Safety Programmes*.

Joint funding scheme

The ODA's Joint Funding Scheme (JFS) for UK-registered charities supports development projects. Relief is specifically excluded; so is research; but some development schemes that have received JFS support may be considered a form of disaster mitigation, including food security projects in Africa and a community shelter project in Latin America that included reconstruction after earthquakes. However, any proposals for disaster mitigation work would have to be placed firmly within a development programme.

Detailed, helpful guidelines are available from:

> NGO Unit
> Room AH 254, Overseas Development Administration, Abercrombie House, Eaglesham Rd, East Kilbride, Glasgow, G75 8EA, UK.
> Tel: 01355 844 000

Other official sources

British Council

The British Council either through its own funds or those it administers on behalf of the British Government and international development agencies, supports the exchange of persons, expertise and experience between Britain and overseas countries. Funds are used to support:

- Training in the UK;

- Attendance at conferences and seminars in the UK;

- Academic and professional links;

- Overseas visits by British specialists;

- South/south exchange when this is developmentally appropriate.

Information regarding The British Council's work can be obtained from:

10 Spring Gardens, London, SW1A 2BN, UK.
Tel: 0171 930 8466 ;   Fax 0171 839 6347

Economic and social research council (ESRC)

The ESRC funds research in the following ways:

- It awards research grants of up to £750 000 for individual projects in response to proposals;

- It supports research centres selected by an annual competition;

- It commissions research within priority areas. Global environmental change is a current interest. Calls for applications are advertised in the national press from time to time;

- It supports the development of research infrastructure for the social sciences, again, advertising in the national press.

Research grants are available to UK universities, polytechnics and colleges of higher education. They can also be given to 'independent research institutes': these can be charities, trusts or companies; and research need not be their main activity. Independent research institutes must get on the ESRC's approved list before applying for a grant, and there is a formal procedure for doing this.

The ESRC has given funds to organisations involved in research and advisory activities relating to disasters. For example, it is funding a major international study of the causes and consequences of involuntary mass departures of migrant communities (a project carried out within the Refugee Studies Programme at Oxford).

One of the themes of the Global Environmental Change Programme (Phase IV), now under way, is 'financial institutions, financial markets and the environment' and suggested areas of work under this heading included: 'the role of the insurance and reinsurance industries, and the assumptions which inform actuarial decisions governing the response to "natural" disasters, such as floods and hurricanes'. The deadline for applications for this phase was November 1994.

Information, including the Guide to Research Funding and details of ESRC's current priority research programmes, is available from:

>   Economic and Social Research Council
>   Polaris House, North Star Avenue, Swindon, SN2 1UJ, UK.
>   Tel: 01793 413000

Natural environment research council (NERC)

Like the ESRC, this funds research carried out by universities, technical colleges and other institutions formally recognised by NERC: only specialist scientific investigations are considered. Its grants cover several main themes which include geology, geophysics, oceanography and hydrology. Funds are usually available for up to three years, and applications must be received by set dates in the year.

The NERC also offers several technical facilities to researchers. Details of these and all the rules and regulations are available in a comprehensive booklet, *NERC Research Grants,* that can be obtained by writing to:

>   The Natural Environment Research Council
>   Polaris House, North Star Ave, Swindon, SN2 1EU, UK.
>   Tel: 01793 411500

Questions concerning applications can also be made by phone (01793 11546 for life sciences; 01793 411657 for physical sciences).

Charitable trusts and foundations

General

There are more than 3100 charitable trusts and foundations registered in the UK, giving a total of over £1 billion a year. They range from tiny family trusts handing out a few hundred pounds at most to large professional organisations whose annual grants total millions. This diversity of size is complemented by an extraordinary variety of funding interests, restrictions and procedures.

It can be difficult to discover if a trust is seriously interested in any given area of work, let alone disasters. The information contained in the standard directories (see section (c) below) is only a starting point and can be misleading.

Charity law is now compelling trusts to be more open about their grant-making activities but there is still a common reluctance to divulge information: trusts are already overwhelmed with far more requests for support than they can hope to meet, and most letters of application do not receive an answer.

The larger trusts tend to be easier to approach but prefer to set and follow their own priorities, seeking out agencies and work to support instead of responding to applications. It is common for trusts and foundations to remain loyal to a select group of organisations year after year.

In many cases trust deeds restrict their grant making to specific types of activity or location, but a large number operate under broad guidelines: terms such as 'general charitable purposes'. For the grant seeker there is no alternative to researching each individual trust: one can often identify its interests from its previous grants.

The biggest constraint is that few trusts will fund organisations that are not registered charities. There is also a widespread reluctance to give money to individuals.

In every case it is vital to build up a good working relationship with the trust or foundation in question. This can take years of effort: in raising money from trusts a long-term plan is essential. Single, one-off, approaches do sometimes strike lucky but as a rule organisations with full-time fundraisers, who can work consistently at building relationships with donors, are best equipped to manage such strategies.

Trusts supporting work on disasters

Charitable trusts formed the bulk of the funding organisations sent questionnaires by the audit team. They were selected because their directory entries indicated an interest in disasters. Hardly any of these were prepared to divulge details of their work and it is likely that they are interested in relief rather than preparedness and mitigation.

Of those that did reply, the most important was the Baring Foundation, which gave nearly £631,000 to disaster-related projects in 1993/4 out of a total of well over £8 million awarded in grants. This included part of a three-year grant to Oxfam's Emergencies Unit and funding for overseas exploratory visits by Merlin. Unfortunately, the collapse of Barings plc, on which the Foundation depended for a large proportion of its income, makes it unlikely that there will be much new work funded for some years to come.

Sources of information on trusts and foundations

The standard source book is:

- *Directory of Grant-Making Trusts*. Edited by Anne Villemur. Charities Aid Foundation. 1995. £53.00. The latest edition appeared after the survey questionnaires had been returned.

Two other directories contain more detailed information about the priorities and activities of the largest trusts:

- *A Guide to the Major Trusts, Volume 1: the top 300 trusts*. Edited by Luke FitzHerbert, Susan Forrester and Julio Grau. The Directory of Social Change. 1995. ISBN 1-873860-49-8. £15.95;

- *A Guide to the Major Trusts, Volume 2: 700 further trusts*. Edited by Dave Casson, Paul Brown and John Smyth. The Directory of Social Change. 1995. ISBN 1-873860-4. £15.95.

A short list of trusts and agencies supporting *development* work in general is given in:

- *The Third World Directory: a guide to development organisations, volunteering opportunities and sources of funding*. Edited by Lucy Stubbs. The Directory of Social Change. 1993. ISBN 1-873860-03-X. £9.95.

Charity projects

Charity Projects (see the entry in the directory) is one of the UK's largest grant-making organisations, raising money through the biennial Comic Relief event of Red Nose Day. Red Nose Day 4 brought in £18.4 million in 1993, and early results suggest that the 1995 campaign may bring in a similar total.

As well as raising massive amounts of money, Comic Relief also has mass appeal. Many millions of people take part in fundraising events and watch the BBC television spectacular on Red Nose Day. Comic Relief takes full advantage of its high profile to inform and educate the public about disasters and development.

Comic Relief arose out of the African famine of 1984-5 and two thirds of the money raised is spent on projects in Africa (the rest goes to projects with disadvantaged people in the UK). Charity Projects' Africa Grants Committee consists of specialists in disasters and development from many different disciplines. Only charities registered in the UK are eligible to receive funding.

In the July 1993 - June 1994 financial year Charity Projects allocated more than £7 758 000 to development and emergency work throughout Africa; £1.4 million was on disaster-related projects. Grants range in size from around £2 000 to around £180 000.

Emergency projects assisted included aid to displaced people (Rwanda), public health (Angola), purchase of grain (Ghana) and water supplies for Sudanese refugees (Uganda).

According to the most recent guidelines, issued in 1993, in its support for emergencies and relief work Charity Projects particularly welcomes projects that:

- Identify means to establish local long-term security in food, health and other civilian needs;

- Cover disaster and emergency preparedness work, research related to emergency preparedness work, and the stockpiling of emergency items;

- Seek to work in chronic situations affecting slum dwellers, long-term refugee and displaced communities or pastoralists.

Among the many kinds of disaster-related development supported by Charity Projects in 1993/4 were projects on community education about landmines, tracing families, dryland agriculture, forestry and livestock, water and sanitation, and health and nutrition.

All applications, whether for emergencies or development, are expected to help build African organisations, respond to local needs, target disadvantaged groups, have a long-term view, measure their impact and learn from their experiences.

Funding policies for Africa are currently under review. Charity Projects is holding discussions on this with a number of NGOs. Greater involvement in disaster prevention is one of the options under discussion - but it is one of many. Funding guidelines will be available after June 1995 (contact the Grants Department on 0171 436 1122).

Other funders

NGOs

Some operational NGOs are also grant makers, the most significant such as Oxfam and Christian Aid being multi-million pound organisations which run their own emergency projects, support local NGOs, and may need consultants for technical assistance, studies and evaluations. Other NGOs, on a smaller scale, have similar aims and act in similar ways.

The diversity of NGOs' aims, activities and operating structures makes this a particularly difficult and complex subject for investigation. The survey and discussions gathered anecdotal and impressionistic evidence which really forms only a starting point for more extensive enquiry.

Within NGOs, as with government and other agencies, disasters and development are usually separated institutionally; at least, at head office level, though in the field a more pragmatic, flexible attitude is likely. Grants to grassroots NGOs working in drought- or flood-prone areas may often in effect be for disaster mitigation and preparedness but will be made through development programmes. This makes it quite impossible to assess how much money might be available for protecting communities against disasters.

There may be opportunities for UK specialists to work with large NGOs, assisting the NGOs' own emergency or development programmes; but it is unlikely that funds will be available for projects outside these unless there is an obvious and direct benefit to the funding NGO. Paid consultancies seem to offer the best prospects.

Some of the larger NGOs may give core funds to disaster institutions in the UK. Oxfam and CAFOD, for instance, have supported the Refugee Studies Programme in Oxford.

Companies

Corporate giving in the UK is on the increase although the levels of funding and strategic planning have a long way to go to catch up with practice in the United States.

Funding from companies in the UK is likely to be on an ad hoc basis in response to requests. Some have well developed community programmes but tend to focus on work locally, or at least, within Britain. Support for work overseas is limited.

All, especially the major corporations, are besieged by supplicants, and very few applications will succeed. Unless there is a good reason for believing that a particular company is likely to favour disaster mitigation work, applications are likely to prove a waste of time. Personal contacts at high level can be invaluable here.

There have been occasional successes. For example, Shell UK, which has a well developed community giving programme, funded research on famine in the mid-1980s.

Sources of information:

- A *Guide to Company Giving*. Edited by Michael Eastwood. The Directory of Social Change. 1993. ISBN 0-907164-96-X. £14.95;
- *The Major Companies Guide*. Edited by David Casson. The Directory of Social Change. 1994. ISBN 1-873860-22-6. £14.95.

The European Union

Although the survey confined itself to UK capacity and resources, it seems appropriate to cover funds managed by the European Commission since these not only use funds provided by the British Government's aid programme but also include one of the few budgets anywhere specifically for preparedness and mitigation.

The European Union gives massive amounts of humanitarian aid[17] - nearly ECU605 million (£465 million @ £1 = ECU 1.3) in 1993, and over ECU 760 million in 1994. Its funding in this area has risen seven-fold in the last four years. Funds are contributed by member states.

The Brussels system is vast, rather bureaucratic and, to the uninitiated, quite bewildering. Even those in the know have to work hard to keep up to date with events. Experience has shown that success at raising money from the Commission depends on making and maintaining good personal contacts with key desk officers; in other words, regular trips to Brussels are essential. It is also worth cultivating good relations with the Commission's delegations in other countries where you plan to run major projects, since they are often required to approve proposals for work.

---

[17] Its definition of humanitarian aid comprises food aid, emergency aid and aid to refugees.

The following paragraphs outline some of the most appropriate sources of funding for disaster preparedness. Where they are known, the names and addresses of useful contacts have been included in the relevant sections. Sections (e) to (g) give additional information on contacts and references.

Over 40 per cent of the European Commission's humanitarian aid is implemented through NGOs. The Commission issues an annual digest describing funds available to NGOs for development and emergency projects as well as giving the names and telephone numbers of the officials responsible.[18] This usually appears in the spring and is well worth having. To obtain a copy try contacting

> Mr Orlando Paleo Labaen
> DG VIII/B/2, G01-1/33, Rue de la Loi 200, B-1049 Brussels, Belgium.
> Tel/Fax: 00 32 2 299 2847

Alternatively, you may be able to obtain one from the Office of the UK Permanent Representative or the NGDO-EC Liaison Committee (see section (e) below).

ECHO

The European Community Humanitarian Office (ECHO) was established in 1992 to bring the Commission's varied humanitarian activities under one roof. Its brief covers emergency aid, prevention and preparedness.

For information about ECHO's work generally, contact:

> Information Section
> Tel: 00 32 2 295 4400; Fax: 00 32 2 95 4572

ECHO's street address is Rue de Genève 3, B-1140 Brussels. For correspondence it is more normal to use the central Commission address at Rue de la Loi 200, B-1049 Brussels.

Emergencies

Most of ECHO's emergency relief allocations are through partner organisations, especially NGOs and United Nations agencies. Agencies receiving such funds sign broad Framework Partnership Agreements in advance that are intended to make the decision-making process swifter and the bureaucracy simpler when disasters occur.[19] Some 150 NGOs and international organisations have now signed. In 1995 six emergency aid budget lines run by ECHO were open to NGOs (details can be found in the Commission's annual digest mentioned above or by contacting ECHO).

Prevention and preparedness

ECHO's prevention and preparedness programme is relatively new and still evolving, and receives only a tiny part of the Commission's total disaster budget: less than 1 per cent of the whole. In 1995, ECU5 million was provisionally set aside for this work (under budget line B7-219N) to cover both ECHO's own activities and its support to other agencies, who may be international organisations, governments or NGOs.

ECHO's programme comprises a range of activities which are arranged under three headings:

- Human resource development (which includes training work for managers, field workers and technicians, some of this to be carried out by regional centres acting as implementing partners);

- Strengthening managerial and institutional capacities (including support to national IDNDR committees in the European Union to promote technology transfer and training programmes in other countries; funding for the United Nations Department of Humanitarian Affairs and the IDNDR Secretariat; and development of national-level preparedness plans, early warning systems and risk mapping);

---

[18] *Digest of Community Resources available for financing NGO Development Activities* - despite the title, emergency work is included.

[19] ECHO is now encouraging recipients of grants for prevention and preparedness to sign these agreements.

- Community-based, low-cost technology projects for disaster preparedness.

The last area appears to be the main route for UK organisations seeking funds for disaster mitigation, and is also interpreted quite flexibly. There is no firm programme here: funding is in response to applications.

Individual grants range between ECU 100 000 and 200 000 and at present are for one year only, albeit with the possibility of renewal. Beneficiaries of the first awards included NGOs, the Pan-American Health Organisation and Organisation of American States. ECHO likes to co-fund projects with other donors though this is not a formal requirement of applications.

The first tranche of preparedness projects was approved late in 1994. 15 projects were funded in this round out of 71 applications submitted. The total value of those supported was nearly ECU 2 225 000. They included research on community-based, low-cost mitigation measures in the Philippines and India, school education for earthquake preparedness in El Salvador and Nicaragua, and creation of emergency credit and flood insurance funds for landless people in Bangladesh. A second round of proposals was due for approval in April 1995.

ECHO expects its support to be made visible by recipients of grants wherever possible and will emphasise this in funding agreements.

Preparedness and mitigation is all managed within one section, ECHO 3, which has recently issued an information booklet and guidelines for applicants. For further information on this programme, contact:

> ECHO 3
> Tel: 00 32 2 295 4615/ 296 9486; Fax: 00 32 2 295 4551

Other directorates general

Despite the reorganisation in 1992, some of the Commission's disaster work remains outside ECHO's control. Four other Directorates General (DGs)[20] are involved in disasters, two focusing on the South.

DG I and DG VIII

DG I, which deals with the Union's external relations as well as development, sometimes addresses disaster problems through its country programmes in Asia and Latin America. It is involved in flood protection work in Bangladesh, for example. It also administers the PHARE scheme for assisting central and eastern European states: part of the PHARE allocation can be spent on humanitarian aid if necessary.

DG VIII is also responsible for development. It runs country programmes in Africa, the Caribbean and the Pacific, and manages long-term food aid operations. Both DG I and DG VIII have substantial refugee programmes. DG VIII handles some emergency aid budget lines that are open to NGOs. It is known to have given financial support to government officials from southern Africa to attend a summer school in the UK.

DG XI and DG XII

DG XI, which deals with the environment, nuclear safety and civil protection, is said to be involved in disasters in Europe; its remit also includes global climate change. DG XII, covering science, research and development, has focused on high-tech research in earth sciences, particularly to prepare against disasters in Europe (see section (c) below).
Consultancies

All DGs need consultants to manage or implement their own operational and research projects. Many invitations are issued to tender for projects as sub-contractors. This work can be very lucrative but requires a solid investment of time and effort in finding out about potential contracts, making oneself known to the relevant officials and getting on mailing lists.

---

[20] There are 23 DGs in Brussels. Each has a different area of responsibility, for instance, transport or energy, and functions like a government ministry. ECHO is unusual in not being part of a DG.

These connections must be maintained, too, for Commission staff often move on to other departments, and many opportunities are not advertised. Notice of major projects is given in the Commission's *Official Journal* which comes out daily, but officials do not need to go to tender for projects of less than ECU 300 000.

To overcome these problems some organisations have set up offices in Brussels, while others hire consultants, often former Commission officials, to go round the corridors on their behalf. The Office of the UK Permanent Representative (see section (e) below) should be able to advise on the best approach here.

Scientific and technological research and development

A few areas of disaster mitigation and preparedness are included in the wide range of activities eligible for support under the European Commission's Framework Programme for Research and Technological Development, managed by DG XII. Framework Programme IV, which runs from 1994 to 1998, is a massive and complex scheme looking to fund innovative work. Natural and man-made hazards feature in the 'Environmental Technologies' theme within the Framework's 'Environment and Climate' component, which itself is only small part of the overall programme.

Projects may last from one to three years. They must be implemented by two partner agencies from two different Member States of the European Union (one may be from one of its Associated States). Participants may be from any kind of organisation, including companies, universities, research institutions or NGOs. Work should benefit European countries.

The Environment and Climate component covers many aspects of the environment, natural resources and climate change. These include research into technologies that forecast, prevent and reduce hydrological, hydrogeological, seismic and volcanic risks, and forest fires. Funds are also available for work on man-made hazards such as pollution and industrial accidents.

There is no official limit on the size of grant available for each project but the total budget for all kinds of environmental technology over the five years is only 120 million ECU and competition for funds is stiff. The Commission usually pays up to 50 per cent of the costs. The application form requires plenty of work and attention to detail. Deadlines are few and far between, and inflexible.

Anybody considering an application should consult the detailed, comprehensive information package on the environment and climate component, which is available from DG XII. The pack also contains application forms and the names and numbers of key contacts in DG XII and Member States who can give additional advice. To obtain it, write to:

>European Commission
>DG XII/D (RTD Actions: Environment), Rue Montoyer 75, B-1040 Brussels, Belgium.

EuronAid

The European Association of Non-Government Organisations for Food Aid and Emergency Relief (EuronAid) aims to provide logistics and financing services to NGOs using Commission food aid in their relief and development programmes. It currently has a membership of 25 European NGOs (UK members are CAFOD, CARE, Christian Aid, Oxfam, Save the Children Fund and Tear Fund). EuronAid is supporting the ODI's Relief and Rehabilitation Network with funds provided by DG VIII.

EuronAid is at:

>Square Ambiorix 10, B-1040 Brussels, Belgium.
>Tel: 00 32 2 732 4696; Fax: 00 32 2 732 4525

Information and contacts in Brussels

Office of the UK permanent representative

For first-timers in Brussels seeking advice and assistance, a visit to the Office of the UK Permanent Representative is recommended. Its job is to represent British interests and help British organisations to secure funding.

Although staff can rarely give you all the information you need (that requires visits to the relevant DGs) they can usually set you off on the right track. One of the most useful services the Office performs is preparing up-to-date lists of desk officers in some of the DGs - invaluable in identifying officials with responsibilities for particular countries or regions.

The office's address is:

> Rond-Point Schumann 6, 6th Floor, B-1040 Brussels, Belgium.
> Tel: 00 32 2 287 8211
> Fax: 00 32 2 287 8398

NGDO-EC liaison committee

The co-ordinating body for national networks with over 800 members, this represents the collective interests of European development NGOs[21] with the Commission in Brussels.

It also provides information on developments in Brussels, including summaries of budget lines and a newsletter. The Liaison Committee has just produced the *NGO handbook*, a new publication to be updated annually, which contains information on budget lines, details of the commission, parliament and other relevant institutions including NGO networks. Free to members of the national platforms; otherwise it costs ECU 15/BF600 (to non-profit organisations) or ECU 30 (other organisations).

Details of this and the Liaison Committee's work generally are available from the head office in Brussels[22] but for full benefits it is necessary to join one of the national networks, which are open only to non-profit organisations.[23]

VOICE

Voluntary Organisations in Co-operation in Emergencies (VOICE) was set up in 1992 as part of the Liaison Committee structure to co-ordinate discussions between NGOs working in emergencies and the Commission, as well as among the NGOs themselves. It provides information to its 65 member NGOs, including a regularly updated directory of humanitarian agencies. Membership costs ECU 1200 a year. For details contact VOICE c/o the Liaison Committee Secretariat (Tel: 00 32 2 732 7137; Fax 00 32 2 732 1934).

Information and contacts in London

European Commission

The Commission maintains an information office at Jean Monet House, 8 Storey's Gate, London SW1P 3AT (0171 973 1992). For any detailed information it makes much more sense to approach staff in Brussels directly.
ODA

---

[21] In European Union vocabulary the term non-government development organization (NGDO) is used to mean NGO.

[22] NGO-EC Liaison Committee, 10 Square Ambiorix, B-1040 Brussels; Tel: 00 32 2 736 4087; Fax 00 32 2 732 1934/32 2 735 0951.

[23] The chair of the UK platform, to whom enquiries should be addressed, is Mike Aaronson at Save the Children (see the section on networks, above).

The ODA's European Community and Food Aid Department may be able to help. The telephone number is 0171 917 0157.

Other books and newsletters

Written information on the EU and the Commission goes out of date very quickly. The European Bookshop in Brussels has the latest editions of all the available guides and directories, and mails lists of new publications.

The bookshop is at:

> Rue de la Loi 244, B-1040 Brussels, Belgium.

Office for orders and correspondence:

> Avenue Albert Jonnar 50, B-1200 Brussels, Belgium.
> Tel: 00 32 2 734 028; Fax: 00 32 2 735 0860

Catalogues and other information on official documents can be obtained from:

> Office for Official Publications of the European Communities
> Rue Mercier 2, L-2985 Luxembourg.
> Tel: 00 352 49928; Fax: 00 352 488573/486817

Commission publications can be bought or subscribed to in the UK through:

> Agency Section
> HMSO Publications Centre, 51 Nine Elms Lane, London, SW8 5DR, UK.
> Tel: 0171 873 9090; Fax: 0171 873 8463

Two helpful books on Brussels and its workings are:

- *The European Community: a guide to the maze*, by Stanley A Budd and Alun Jones (published by Kogan Page) is an excellent, comprehensible introduction to the whole system and the workings of the Commission, and is updated regularly;

- *The European Commission Information Handbook*, which is published by the EC Committee of the American Chamber of Commerce in Belgium[24], contains lists of departments and their roles, with names and telephone numbers of the main officials.

The *Euro-CIDSE News Bulletin* provides frequent (sometimes monthly) news on new developments in Brussels and is very informative.[25] It has information on the annual negotiations between the commission and the European parliament over the aid budget and allocations to individual budget lines, but information about application procedures must be obtained from the relevant Dgs.

Conclusions

Funds Available

At first glance, UK funding prospects for disaster preparedness and mitigation are not rosy. As this survey shows, few donors are interested in the field. The financial resources available specifically for such work are limited, especially when set against the demand from practitioners seeking grants or contracts. Most major funding sources are outside Britain: United Nations and bilateral agencies, national governments, and some overseas NGOs and foundations. These have a particular interest in

---

24 Avenue des Arts 50, BTE 5, B-1040 Brussels; Tel: 00 32 2 513 6892/513 6770; Fax : 00 32 2 513 7928.

25 Contact Eileen Sudworth, EURO-CIDSE Secretariat, Rue Stevin 16, B-1040 Brussels.

subcontracting work, hiring consultants or commissioning research studies, although few of the major government and international agencies make any formal financial provision to support work on disaster mitigation and preparedness.

On the other hand, the disaster field is so broad in its range and so varied in the types of its activity that individuals and organisations seeking funds can approach a wide variety of donor agencies. Rather than seeking 'disaster funds' per se, they can go to funders interested in, for example, scientific and technical research, or in giving bursaries for individuals to attend training courses. Activities in disaster mitigation are sometimes included within long-term development projects or programmes, and have been supported by agencies and funding schemes geared to development.

Effect on disaster organisations

The directory produced with the survey shows the large number of British agencies and individuals working in disasters. All need finance and many are competing for grants or contracts from the same funders, adding to the pressure on scarce resources.

Even when grants are awarded, this only relieves the pressure for a short while. Most of the money available is for single, one-off projects. Few organisations working in disasters ever feel financially secure for a long period of time, especially small commercial and consultancy outfits. They have to continue the chase for new funding and new projects, and are denied sufficient opportunity to reflect on past work and develop longer-term strategies.

The insecurity and competitive fundraising climate make disaster specialists highly possessive of their 'own' funders and suspicious of others trying to secure money from the same sources. Joint activities between applicant organisations to minimise the effort of fundraising and maximise the application of donor support tend therefore to be exceptional, one-off achievements, rarely involving more than two agencies.

Opportunities and recommendations

Unless the funds available increase, the future for the UK disaster community looks likely to remain similar to today. If anything it may worsen: in disasters as in development, new organisations are always entering the scene.

What can be done, then, to mitigate these tendencies and improve prospects? In disaster terminology, what coping strategies should we adopt? Progress could be made in two areas: more collaboration and partnerships; and work to expand existing funding schemes and establish new ones. Advances on both fronts are necessary. Neither will happen without genuine commitment within the disaster community itself.

Partnerships and collaboration

This line of approach offers the best short-term potential for increasing access to funds as well as reducing the cost of raising them.

Individuals and organisations of all kinds have to be prepared to plan and implement work together much more extensively, and often, than they do now. Collaboration should involve not just two but even several enterprises.

In terms of fundraising there are many advantages to be gained from partnerships.

They reduce the sheer effort of competing for funds. The labour of writing proposals can be shared or delegated. If this allows those taking part to spend a bit more time on preparing and refining their proposals, it will not only enhance the chances of success but should also improve the quality of the projects themselves.

Each partner in a team will have its own close contacts within particular funder agencies. This offers the team a larger number of funding opportunities, allowing multiple applications and perhaps covering wider areas of work, with a correspondingly greater likelihood of getting funding. Individual partners can take responsibility for liaising with their 'own' funders on behalf of the team; they need not be afraid of somehow forfeiting their special links.

Voluntary sector agencies and registered charities have a comparative advantage since a number of funding schemes are open only to this type of organisation. But they often need inputs from consultants, academic researchers or other specialists, who may have their own comparative advantage through better links with multilateral agencies or research funders. The advantages of working together are obvious.

Many donors prefer to co-finance projects with other agencies, either because they want to spread their own grants as widely as possible or because they like to be associated with other funders. The team method of project planning and fundraising is more likely to bring about co-financing opportunities than individual efforts.

Linking different kinds of work widens the range of potential funders. Take the example of a research institution setting up a project with a field agency. By including an operational component in the work, the research institution now has access to agencies that would never fund pure research but may consider it as part of fieldwork.

There is now fairly widespread acceptance of the need to incorporate disaster preparedness, prevention and mitigation in development planning. Collaborative projects between organisations working in development and specialising in disasters could not only ensure this happened but also allow access to some of the major funding schemes for development. The current institutional separation and weak contacts between the two fields have been a real brake on progress here.

Given that the disaster community in the UK is not noted for its culture of co-operation, a number of historical and psychological hurdles will have to be overcome to bring about greater collaboration. Clearly, this will take time.

We must also accept that agencies have their own different funding needs and pressures. A large charity with an extensive public fundraising programme can afford time to plan and develop its own long-term work - and then look for project funding. That is a luxury that organisations of another kind or size cannot afford: they may be compelled to rush for funds when they become available, tailoring their project proposals to the funder's interests and requirements.

Nevertheless, the attempt to work together must be made. Only a few enthusiasts are needed to set the process in motion.

Expansion and creation of funding schemes

This second line of approach is more complicated and long-term. It requires concerted effort to convince donors of the need for disaster preparedness and mitigation.

Better publicity is a prime need. Disaster agencies are often highly effective lobbyists for themselves and their cause on a one-to-one basis with key individuals in donor organisations. For the long term a broader strategy is needed to inform and influence donors. This will require producing the whole apparatus of reports, evaluations, leaflets, books, articles, seminars, exhibitions, videos and other means of communication that explain the issues and publicise success stories.

Few organisations can afford to carry out this kind of work on their own. The IDNDR and its associated structures are beginning to help raise awareness of mitigation. It is bound to be a slow process but perhaps more could be done, and perhaps there could be a greater focus on influencing donors of all kinds.

Meanwhile, the entire UK disaster community can put pressure on donors currently backing disaster mitigation by overloading them with applications! This may not force them to increase their mitigation and preparedness budgets but it is a clear indication to policy makers that the supply of funds is not meeting the demand.

Information Sources

Libraries and information centres

The questionnaire asked respondents to indicate which libraries/information centres, inquiry services and photo, video and film library services they provided *that were accessible for outside use*. The response from the questionnaires state that 29 per cent of organisations (including individuals) indicated yes to having developed a library/information centre. Not all of them are available for public use. 15 per cent of the respondents had a photographic library/video/ slide collection and 23 per cent offered an inquiry service, although the audit did not identify what the nature of the inquiry services are.

Of the organisational categories 41 per cent of NGOs have libraries/information centres whilst 45 per cent of academic/research bodies have this resource. 26 per cent of consultancies have a library/ information centre whilst only 10 per cent of individuals have a collection of literature. Probably the best information centres, to which access is usually given on request, currently include:

- The Food Studies Group (FSG) at the University of Oxford;
- The Institute of Development Studies (IDS) at the University of Sussex;
- The International Institute for Environment and Development (IIED);
- The Overseas Development Institute (ODI) at Regent's College, London;
- The Refugee Studies Programme (RSP) at the University of Oxford.

Information on forthcoming publications regarding disaster mitigation and preparedness is often publicised through the IDNDR Project Office 'new publications' leaflet. The leaflet for September 1995, new publications on disaster mitigation publicises two new publications: *Megacities: Reducing Vulnerability to Natural Disasters* (Thomas Telford Services Ltd, £30), and *Structures to Withstand Disaster* Thomas Telford Services Ltd, 1995, £30). The IDNDR address is:

IDNDR Project Office
The Institution of Civil Engineers, 1-7 Great George Street, London, SW1P 3AA, UK.
Tel: 0171 839 9963/4; Fax: 0171 233 1806.

Bookshops

The number of specialised bookshops and publishers remains very limited, with disaster mitigation and preparedness remaining a section within the larger development context. Hence books on this subject can be found in generalist bookshops, often within geography and development departments. Most organisations that produce literature for sale have a publications list and it seems that this sector has developed as an offshoot to the organisation's main work as a profit venture and for the dissemination of information. Particular *specialist* bookshops include:

- Africa Book Centre Ltd, 38 Kings Street, London, WC2E 8JT, Tel: 0171 240 6649
- Oxfam Resources Centre, Oxfam, 217 Banbury Road, Oxford, OX2 7DZ, Tel: 01865 311311
- Intermediate Technology Bookshop, 103/5 Southampton Row, London, WC1B 4HH,
  Tel: 0171 436 976, Fax: 0171 436 2013
- Thomas Telford Bookshop, ICE, 1-7 Greta George Street, London, SW1P 3AA

Outside the UK, European sources of official documents include:

- European Union Bookshop, Rue de la Loi 244, B-1040 Brussels, Belgium.

    Office for orders and correspondence: Avenue Albert Jonnar 50, B-1200 Brussels, Belgium.
    Tel: 00 32 2 734 028; Fax: 00 32 2 735 0860

Catalogues and other information on official documents can be obtained from:

- Office for Official Publications of the European Communities, Rue Mercier 2, L-2985 Luxembourg.
    Tel: 00 352 49928; Fax: 00 352 488573/486817

Commission publications can be bought or subscribed to in the UK through:

- Agency Section, HMSO Publications Centre, 51 Nine Elms Lane, London, SW8 5DR, UK.
  Tel: 0171 873 9090; Fax: 0171 873 8463

Journals

The five publications identified in the audit questionnaire were *Hazards Forum, Disaster Management, Disasters Journal, Civil Protection* and *Emergency*. Additional space for respondees to list any other journals not mentioned was provided. Added journals included *Mapping Europe, GIS Europe, STOP Disasters, The Journal of Refugee Studies* and the *Relief and Rehabilitation Network*.

The responses given to the listed journals were surprisingly low: 9 per cent to *Hazards Forum*, 7 per cent to *Disaster Management*, 19 per cent to *Disasters*, 7 per cent to *Civil Protection* and 3 per cent to *Emergency*. However these findings represent *subscribers*, not *readers*, and no distinction has been made to size of organisation; hence for example 1an organisation of say 100 staff might only account for one entry.

The largest recorded journal subscribed to, *Disasters*, is published four times a year, in March, June, September and December. *Disasters* aims to bring together research on disasters and relief and emergency management. Its aim is to promote the interchange of ideas and experiences between relief practitioners and academics. It includes field reports, case studies, articles of general interest and academic papers. *Disasters* is published by Blackwell Publishers. Their address is:

Blackwell Publishers, 108 Cowley Road, Oxford, OX4 IJF, UK.

It is edited by the Overseas Development Institute. Books for review should be sent to:

The Editor, ODI, Regent's College, London, NW1 4NS, UK.

Databases

It is unknown how many current databases of disaster information exist. However it is safe to say there are very probably many, although those accessible to the public domain is limited: information carries high (financial) value. Some of the most extensive databases exist within the insurance industry, for example *Expose*, Alexander Howden's catastrophe mapping model and information source. The database contains extensive information on vulnerability and past and potential windstorm damage. Other databases held with funders may contain information on organisations and on past projects However these also are largely inaccessible.

RiskMap

in 1992 save the Children (SCF) began developing a database and model for risk mapping for food security, intended primarily as a contribution to improved famine early warning and food aid decision-making. RiskMap offers the user 'hands-on' examination of the basic way people feed themselves in the different parts of a given country, and 'hands-on' imposition by the user of the known elements of a current or past disaster situation, natural or man-made. This leads to a mapped and graphed result which gives a best estimate of the proportion of a population likely to be in food deficit, and the degree of that deficit.

The computer model is based on an analysis of the rural family's usual access to food seen in terms of income: direct food production or gathering, and cash income for food purchases. This is further set in the context of local markets. The user sets a disaster problem onto this database, specifying what is known at a given time of crop and grazing losses, and/or disruption access to given markets for food purchase, livestock and cash-crop sales, or employment. The programme then runs through a series of steps towards the final result output, taking into account the consumption of food stocks, use of cash and capital assets, and redistribution from the richer to the poorer in the community.

The SCF RiskMap is developed in association with UN Food and Agricultural Organisation with European Community funding. Refinement of the programme and extension of the country databases continues in 10995. Further details can be gained from:

Penny Allen, Risk Mapping Adviser
Save the Children, Mary Datchelor House, Grove Lane, London, SE5 8RD, UK.
Tel: 0171 703 5400; Fax: 0171 793 7626; E Mail 100144.1457@compuserve.com

EM-DAT

The Centre for Research on the Epidemiology of Disasters (CRED), at the university of Luvain, Brussels has developed a system of databases for global disaster management. The system, called EM-DAT[26], 'draws on its existing disaster information, information network and computer system' (*World Disasters Report* 1995, IFRCRCS, Page 93). There are now over 10 000 records of disaster events from 1900 available The criteria for entry is ten or more deaths and/or 100 affected and/or an appeal for assistance.

Further information can be found the World Disasters Report 1995, International Federation of Red Cross and Red Crescent Societies, Nijhoff, Switzerland, 1995 where EM-DAT is described in detail. Otherwise the address for CRED is:

Centre for Research on the Epidemiology of Disasters (CRED)
Catholic University of Louvain, School of Public Health, Clos Chapelle-aux-Champs 30-34
1200 Brussels, Belgium.
Tel: 32 2 764 3327/764 3823; Fax: 32 2 764 3328.

The audit database

A key component of this audit was the development of a database of information. The database contains all the information represented in this publication. It contains each of the main sections of the audit and allows for cross interrogation of the following categories: organisations, regions of activity, hazard expertise, work content and skills. There were no restrictions on information entered into the database, other than that the information followed the key theme of disaster mitigation, preparedness and (eventually widened to) response. Further information about the audit database can be obtained by contacting:

The Oxford Centre for Disaster Studies PO Box 137 Oxford, OX4 1BB, UK.
Tel: 01865 202772; Fax: 01865 202848.

The Internet

The Internet[27] is unquestionably the fastest growing global medium for rapid access to information and communication. Its use in the area of disaster mitigation and preparedness is extensive: in terms of rapid information gathering from all around the world (detailed disaster statistics of the Kobe earthquake were available on the Internet within several hours of the disaster), the exchange of ideas and networking through discussion groups, the access to large amounts of detailed information through Internet library searches, and the ease and cost effectiveness of global communication via
E-Mail.

The number of Internet users, and hence the amount of information available, is growing at an almost exponential rate. The amount of relevant information is almost infinite (governmental and intergovernmental country reports, satellite images, disaster statistics, etc), as are the growing number of discussion groups and new *home pages* (contents pages for organisations, institutions and specific types of information).
Information on the Internet is accessed through the use of dedicated information retrieval software such as *Archie* (which finds information that is stored on computers around the world) and *Gopher* (software

---

[27] Background information gained from Dr Robin Stephenson and AT Brief no.10, *Electronic Mail and the Internet*, Appropriate Technology, Volume 21, Number 3 December 1994

that organises information into a series of menus, allowing interrogations to become more specific as successive menus are accessed). The World-Wide Web combines images, text and sound to allow users to access information stored on *Web servers*.

The following list comprises some of the key Internet addresses currently in use for accessing information related to disasters.

- http://www.disaster.org

- http://adder.colorado.edu/hazctr/home.html

- http://hoshi.cic.sfu.ca/hazard/
  for access to the HazardNet home page

- http://hoshi.cic.sfu.ca/hazard/idndr.html
  for access to the IDNDR home page

- http://www.vita.org

- http://www.fema.gov

- http://www.info.usaid.gov

- http://tin.ssc.plym.ac.uk/gemc.html
  for access to the Global Emergency Management Disaster Councelling Support Network

- http://www.foe.co.uk
  the home page for the Friend of the Earcth, providing information on navigating the Internet as well as environmental information

Further information regarding the Internet can be found in:

- Gilster, Paul, *The Internet Navigator*, John Wiley and Sons, UK and USA, 1992.
- Krol, Ed, *The Whole Internet Catalog and User's Guide*, O'Reilly and Associates, USA, 1992
- Lane, Graham, *Communications for Progress: A Guide to International E-Mail*, CIIR, London, 1990

Summary of organisational activities

The following table summarises the activities of organisations according to hazard, region and skills as recorded on the returned questionnaires.

Organisation	Hazard	Region	Skills

Organisations:
1. Action Health
2. Action Water Charity
3. ActionAid
4. Adventist Development and Relief Agency (ADRA)
5. Afghanaid
6. Agency for Research & Coop'n in Devel't (ACORD)
7. Agricultural Extension and Rural Development Dept
8. Aids, Care and Training (ACET)
9. British Association for Immediate Care
10. British Consultants Bureau
11. British Geological Survey
12. British Red Cross Society Headquarters
13. Building and Social Housing Foundation
14. Cambridge Architectural Research Ltd
15. Cargil Attwood Consultants
16. Catastrophe Reinsurance
17. Catholic Fund for Overseas Development (CAFOD)
18. Centre for African Studies, University of Cambridge
19. Centre for Arid Zone Studies, University of Wales
20. Centre for Devel't and Emerg'y Planning (CENDEP)
21. Centre for Environ'l and Human Settlements (CEHS)
22. Centre for International Health
23. Centre for the Study of African Economies
24. Centre for Urban & Regional Studies, Univ'y of B'ham
25. Charity Projects (Comic Relief)
26. Christian Aid
27. Climatic Research Unit School of Env'l Science

Attribute	Organisations (numbered as above)
Hazard	
Drought	4, 5, 6, 7, 11, 12, 14, 17, 19, 20, 25, 26
Hurricane and Cyclone	4, 5, 11, 12, 14, 15, 16, 17, 25, 26
Famine	4, 5, 6, 12, 14, 17, 19, 20, 25, 26
Flood	4, 5, 11, 12, 14, 17, 19, 20, 25, 26
Landslide	10, 11, 12, 14, 15, 16, 17, 19, 25, 26
Earthquake	10, 11, 12, 14, 15, 16, 17, 19, 25, 26, 27
Volcano	10, 11, 12, 14, 15, 16, 17, 19, 25
Complex emergency	12, 15, 17, 18, 19, 22, 23, 25, 26
Disease and epidemic	12, 17, 19, 21, 24, 25, 26
Region	
Latin America inc Mexico	4, 10, 11, 12, 17, 19, 25, 26
Australia and Pacific	4, 11, 12, 19, 25
The Caribbean	10, 11, 12, 14, 17, 19, 25, 26
Western Europe	4, 10, 11, 12, 14, 17, 19, 21, 25
E Europe/Former Soviet Union	10, 11, 12, 14, 17, 18, 19, 25, 26
South/South East Asia	10, 11, 12, 14, 15, 16, 17, 18, 19, 22, 25, 26, 27
East Asia	10, 11, 12, 14, 17, 19, 25, 26
Middle East	10, 11, 12, 17, 19, 22, 23, 25, 26
Africa	10, 11, 12, 14, 17, 18, 19, 20, 22, 23, 24, 25, 26
USA and Canada	11, 17
Skills	
Food Security	8, 9, 10, 11, 12, 17, 19, 20, 25, 26
Engineering	10, 11, 13, 14, 15, 17, 22
Seismology	10, 11, 13, 14, 15
Building and architecture	10, 13, 14, 15, 17, 20, 22
Physical planning	10, 13, 14, 15, 17
Health/epidemiology/nutrition	8, 10, 17, 18, 20, 22, 27
Agriculture	7, 8, 10, 17, 18, 19, 20, 26, 27
Forestry	10, 17, 18, 19, 20
Hydrology	11, 14, 17, 19, 27
Geomorphology	11, 14
Volcanology	11, 14
Anthropology	14, 17, 19, 20
Conflict prevent'n/tension rec	14, 17, 19, 20, 22, 25
Meteorology	27
Insurance/reinsurance	15, 16
AT/indigenous knowledge	14, 17, 19, 22, 25, 26
Energy	11, 17
Remote sensing	11
Development economics	17, 19
Transport	10, 14, 15, 17, 19, 22, 23, 24
Training	6, 10, 11, 14, 15, 17, 19, 20, 22, 23, 24, 25, 26
Information management	10, 11, 14, 15, 17, 19, 22, 23
Technical research	10, 11, 14, 17, 19
Social science research	17, 18, 20, 21
Communications	10, 11, 17, 25
Media	10, 18, 19, 25, 26

Organization					
Concern Worldwide	o		o	o	
Cranfield Disaster Preparedness Centre (CDPC)	o	o	o	o	o
Crown Agents	o	o	o	o	o
Datum International	o	o	o	o	o
Dept of Civil Engineering, Imperial College	o	o	o	o	
Dept of Geography, University of Liverpool	o	o	o	o	
Dept of Civil Engineering, University of Portsmouth	o	o	o	o	
Dept of Geography, University of Cambridge	o	o	o	o	
Dept of Mathematics, University College, London	o	o			
Dept of Social Anthropology, Univ'y of Manchester	o	o	o	o	o
Department of the Environment (DoE)	o	o	o		
Dept of Community Medicine, Cambridge	o	o	o	o	
Dept of Geography, Chester College	o	o	o	o	
Earthquake Engineering Research Centre	o	o	o	o	o
ECHO International Health Service	o	o	o	o	o
Entec Europe Ltd	o	o	o	o	o
ETC (UK)	o	o	o	o	o
Farm Africa	o				o
Flood Hazard Research Centre	o	o	o	o	
Food Studies Group	o	o	o		o
Geology Dept, University of Bristol	o	o	o	o	
Gifford and Partners	o	o			o
Green Cross UK		o	o		
Hadley Centre Meteorological Office	o	o	o		
Harvest Help	o	o	o	o	
Hazard and Risk Management Studies (HARMS)	o	o	o		o
Health & Nutritional Status Advisory Unit (HANSA)	o	o	o		
Health Projects Abroad	o	o	o	o	
HelpAge International	o	o	o	o	o
Homeless International	o	o	o	o	o
Institute of Child Health, University of London	o	o	o	o	
Institute of Civil Defense and Disaster Studies	o	o	o	o	o
Institute of Development Studies, Sussex	o	o	o	o	o
Institute of Terrestrial Ecology	o	o		o	o
Intermediate Technology	o	o	o	o	o
International Christian Relief	o	o	o	o	o
International Extension College	o	o			o
Intern'l Institute for Environment and Devel't (IIED)	o	o	o	o	o

Organisation	Hazard									Region									Skills																							
	Drought	Hurricane and Cyclone	Famine	Flood	Landslide	Earthquake	Volcano	Complex emergency	Disease and epidemic	Latin America inc Mexico	Australia and Pacific	The Caribbean	Western Europe	E Europe/Former Soviet Union	South/South East Asia	East Asia	Middle East	Africa	USA and Canada	Food Security	Engineering	Seismology	Building and architecture	Physical planning	Health/epidemiology/nutrition	Agriculture	Forestry	Hydrology	Geomorphology	Volcanology	Anthropology	Conflict prevent'n/tension red	Meteorology	Insurance/reinsurance	AT/Indigenous knowledge	Energy	Remote sensing	Development economics	Transport	Training	Information management	Technical research	Social science research	Communications	Media
Intern'l NGO Training and Research Centre (INTRAC)	o							o						o				o													o									o			o		
International Seismological Centre		o				o																o																							
Karuna Trust			o	o	o										o																														
Living Earth								o				o	o					o									o	o	o	o	o	o	o		o	o		o	o	o	o	o	o	o	o
Llewelyn Davies Planning					o	o															o																								
Lloyd's Register (Civil and Structural Eng.)		o				o															o	o																							
London School of Hygiene & Tropical Medicine	o		o					o	o	o		o		o	o	o	o	o	o						o																		o		
Marie Stopes International									o																o																				
Medical Emergency Relief Internationcl (MERLIN)								o	o					o	o		o	o							o							o								o	o		o	o	o
Mott MacDonald Group				o		o							o				o	o			o		o					o								o			o						
Multilateral Research Economics Department (ODA)	o	o	o	o	o			o	o	o	o	o		o	o	o	o	o		o						o	o	o	o		o	o			o			o		o	o	o	o	o	o
NERC Unit for Thematic Information Systems (NUTIS)	o	o	o	o	o									o				o																			o								
Overseas Development Administration (ODA)	o	o	o	o	o		o	o	o			o	o	o	o	o	o	o								o	o	o	o	o	o	o						o		o	o	o	o	o	o
Overseas Development Institute (ODI)	o	o	o	o	o		o	o	o	o	o	o	o	o	o	o	o	o	o	o				o		o	o	o	o		o	o			o			o		o	o	o	o	o	o
Oxfam	o	o	o	o	o		o	o	o		o				o	o	o	o	o	o						o	o					o								o	o	o	o	o	o
Oxford Centre for Disaster Studies (OCDS)	o	o	o	o	o		o			o		o	o	o			o	o						o							o														
Quaker Peace and Service								o										o														o											o		
Rural Resources Management Ltd.										o	o	o						o		o	o					o	o	o			o	o			o		o	o		o	o	o	o	o	o
SAFE (Support Action for Emergencies)	o		o	o	o	o	o	o	o	o	o	o			o	o	o	o		o						o	o				o	o													
Saferworld	o							o		o		o						o													o	o			o			o		o	o	o	o	o	o
Save the Children Fund	o	o	o	o	o	o		o	o	o	o	o		o	o	o	o	o		o																		o		o	o	o	o	o	o
School of Devel't Studies, University of East Anglia	o	o	o	o	o					o					o	o	o	o		o						o	o	o			o				o			o		o	o	o	o	o	o
School of Humanities, University of Greenwich	o							o	o						o		o	o																			o			o	o	o	o	o	o
School of Oriental and African Studies (SOAS)	o					o											o	o																											
Scottish Catholic International Aid Fund (SCIAF)	o					o	o	o							o			o								o																			
Sedgwick Global																																		o		o									
Silsoe Research Institute	o		o															o		o	o					o	o	o			o				o					o	o	o			
SOS Sahel	o		o															o		o						o	o	o				o			o										

- South Bank University
- Systems Group, The Open University
- Tear Fund
- The Centre for Crisis Psychology
- The Research Centre, University of Luton
- Trocaire
- UK Committee for UNICEF
- UK- Med
- United Kingdom Jewish Aid
- United Nations Association (UNA)
- Volcano Geophysics Group
- Water Engineering and Devel't Centre (WEDC)
- Wind Engineering Society
- World Assoc'n for Disaster and Emergency Medicine
- World Conservation Monitoring Centre
- World Vision UK

Current activities

The following list is compiled from the information provided from the completed questionnaires. It gives details of current projects undertaken by both organisations and individuals. Any gaps in the tables reflects information not given by respondents.

Current activities

The following list comprises information returned on the questionnaires relating to activities *ongoing or due to begin in 1995*.
Questions were asked regarding title, location, duration and brief objectives. Where gaps occur information was returned incomplete

Organisation and Contact		Location	Project	Objectives	Duration
ActionAid	Margie Buchanan-Smith	Zaire	Assisting Rwandan refugees		Since 1994
ActionAid	Margie Buchanan-Smith	Ghana	Response to drought and floods	Food security and community grain banks.	Since 1994
ActionAid	Margie Buchanan-Smith	Ethiopia	Food security	Grain distribution.	Since 1994
Action Water Charity	Neil Battersby	Goma Rwanda	Provision of water tankers, hoses and pumping equipment		
Adventist Development and Relief Agency (ADRA)	John Arthur	Goma	Rwanda relief	1) Water supply 2) Clothing and plastic sheeting etc. 3) Establishment of a small hospital.	
Adventist Development and Relief Agency (ADRA)	John Arthur	Sarajevo, Mostar, Belgrade, Split, Zagreb	Relief	1) Supplies of humanitarian aid 2) Delivery of 320 000 letters and 290 000 family food parcels to Sarajevo.	
Afghanaid	Wynn Flaten	Jalalabad	Emergency shelter programme	To provide shelter to vulnerable families.	Mid 1994 - 95
Afghanaid	Wynn Flaten	Badakhstan	Emergency food programme	To provide emergency food relief to vulnerable groups.	Late 1994 - 95
Afghanaid	Wynn Flaten	Kabul	Water and sanitation programme	To provide safe water and sewage infrastructure.	Mid 1994 - 95
Agency for Research and Co-operation in Development (ACORD)	Peter James	Mogadishu Somalia	Rural development and rehabilitation	Institutional development with government organisations. Stronger social fabric to withstand future conflict.	
Agency for Research and Co-operation in Development (ACORD)	Peter James	Rwanda	Rehabilitation and emergency relief	Relief to communities displaced by civil war.	May 1994 - Dec 95

Organisation and Contact		Location	Project	Objectives	Duration
Aids, Care and Training (ACET)	Maurice Adams	Thailand	Thailand AIDS care	Initiating community care programmes for those with AIDS.	1991 - present
Aids, Care and Training (ACET)	Maurice Adams	Throughout Romania	Romanian professional AIDS training	With UNICEF and Gat to reduce HIV in communities and hospitals.	1991 - present
British Geological Survey	Chris Browitt	Basiucasa Italy	Rainfall induced landslides	Use of GIS for storing and manipulating data.	
British Red Cross Society Headquarters	Mike Adamson	Ethiopia	Food distribution and long term development		Since 1994
British Red Cross Society Headquarters	Mike Adamson	Former Yugoslavia	Emergency relief	Water support programmes, food parcels, soup kitchens and tracing families.	Since 1994
British Red Cross Society Headquarters	Mike Adamson	Cambodia	Medical support	Helping rebuild a medical infrastructure, tracing families and medical support.	Since 1994
Centre for Development and Emergency Planning (CENDEP)	Hugo Slim	Bangladesh and Kenya	Disaster management training workshop	NGO Training.	
Centre for Development and Environmental planning (CENDEP)	Hugo Slim	Oxford	Peacekeeping research	To asses the social impact of peacekeeping.	
Centre for Urban & Regional Studies	Mark Duffield	South Sudan	Review of Sudan Emergency Operation Consortium (SEC)	Programme review.	
Christian Aid	Jenny Borden	Mahrashtra India	Relief and rehabilitation	Housing provision.	
Christian Aid	Jenny Borden	Sierra Leone	Relief and rehabilitation	Provision of assistance.	
Christian Aid	Jenny Borden	Southern Sudan, Angloa, Bosnia	Relief and rehabilitation	Provision of assistance.	
Christian Aid	Jenny Borden	Rwanda	Relief and rehabilitation	Provision of assistance.	Long term
Christian Aid	Jenny Borden	Bangladesh	Relief and rehabilitation	Cyclone shelters.	

Organisation and Contact		Location	Project	Objectives	Duration
Concern Worldwide	Nick Guttmann	Rwanda, Burundi, Tanzania, Zaire, Angola, Mozambique	Emergency relief programmes		
Cranfield Disaster Preparedness Centre (CDPC)	Ken Westgate	Africa	UN disaster management training programme	Disaster management.	Ongoing
Cranfield Disaster Preparedness Centre (CDPC)	Ken Westgate	Africa	Training materials for indigenous NGOs in Africa	To provide training packages for indigenous NGOs in Africa.	Ongoing
Cranfield Disaster Preparedness Centre (CDPC)	Ken Westgate, Mary Myers	Mali, Burkina Faso, Eritrea	Developing the use of local radio in disaster preparedness	To investigate and promote the use of local radio in disaster preparedness.	
Crown Agents	Alan Matthews	Bosnia	Emergency aid convoys	Operation of UN humanitarian relief convoys.	1992 to date
Crown Agents	Alan Matthews	Rwanda	Tribal genocide	Assessment, advisory and convoy operations.	1994 to date
Crown Agents	Alan Matthews	Zambia	Drought relief	Provision of drilling equipment.	
Dept of Civil Engineering Imperial College	C J Kerr	Oman and Middle East	Assessment of potential for flooding in arid areas	To assess the hydrology of the areas and the impact of storm-runoff on structures etc.	
Dept of Civil Engineering Imperial	C J Kerr	Loma Prieta, Kakumata, Greece, Armenia	Fieldwork at earthquake sites	To review damage and assess buildings for safety.	1 week
Dept of Geography University of Liverpool	David Chester	Italy	Volcanology applied	Hazard reduction.	
Dept of Geography University of Liverpool	David Chester	Azores	Volcanology applied	Hazard reduction.	
Dept of Civil Engineering University of Portsmouth	Prof B E Lee	University of Portsmouth and Colorado State University	Shelter effects for low rise building groups	To determine wind shelter available from low rise buildings.	1995 - 1998

Audit of UK Assets in Disaster Mitigation, Preparedness and Response

Organisation and Contact	Location	Project	Objectives	Duration
Dept of Civil Engineering University of Portsmouth — Prof B E Lee	Aerodynamics R&D Centre, Sichan Province China	Dynamic wind loads on tall buildings	A parametric study of design variables which influence dynamic wind loads.	1991-1996
Dept of Geography University of Cambridge — Clive Oppenheimer	Kamchatka, Russia	Volcanoes in Kamchatka	Remote surveillance of volcanoes.	1995 - 1998
Dept of Geography University of Cambridge — Clive Oppenheimer	Jarva	Volcanic hazard in Indonesia	Hazard volcano monitoring.	Ongoing
Dept of Geography University of Cambridge — Clive Oppenheimer	Global	Remote sensing of volcanoes	Development, testing & implementation of remote sensing system for volcano monitoring.	
Dept of Community Medicine — Peter Baxter	Cuba	ODA UK scientific collaboration in epidemic of nempathy	Investigation epidemic of nempathy.	2 years
Dept of Community Medicine — Peter Baxter	Irazu Volcano Costa Rica		Disaster preparedness.	
Dept of Community Medicine — Peter Baxter	Poas Volcano Costa Rica	European Community scientific co-operation	Air pollution and health disaster preparedness.	
Dept of Community Medicine — Peter Baxter	Furanj Volcano Azares	European Community volcano project	Risk assessment/disaster preparedness.	Ongoing
Dept of Geography Chester College — Martin Degg	Chester UK	The use of Geographical Information Systems (GIS) in landslide hazard mapping	Explore the role of GIS in landslide hazard and risk mapping.	1993 - 96
Dept of Geography Chester College — Martin Degg	Chester UK	Earthquake hazard in Egypt: community perception and response	In the wake of the 1992 "Cairo earthquake" to examine perception of response, earthquake hazard rd at community level.	1994 - 97
Dept of Geography University of Durham — Ewan Anderson	Dubai, Abu Dhabi, Durham	Worldwide Centre	1) Telecommunications and database 2) Early warning 3) Research into communities at risk.	

Organisation and Contact		Location	Project	Objectives	Duration
Dept of Geography University of Durham	Ewan Anderson	Bombay	Community health care	To improve water related services in slums.	
Dept of Geography University of Durham	Ewan Anderson	Bombay	Disaster relief centre	Establish centre.	
Development Planning Unit University of London	Babar Mumtaz	Global	Disaster mitigation manual for megacities	Inputs into ODA funded manual being prepared by Institute of Civil Engineers.	
Earth Observation Sciences Ltd	D R Sloggett	European waters	Oil spill detection	Automate the detection of oil.	2 years
Earth Observation Sciences Ltd	D R Sloggett	Europe and Africa	Forest fire	Automate the detection of forest fires in tropical vegetation and in Europe.	1 year
Earth Resources Centre	John Merefield	Venezuela	Soil gas geochemistry	For seismic risk.	1994 - 96
Earth Resources Centre	John Merefield	Reunion Island	Water gas geochemistry	In volcanic risk.	1995
Earth Resources Centre	John Merefield	Erzurum Turkey	Soil gas geochemistry	For seismic risk.	1991 - 95
Earthquake Engineering Research Centre	Prof R T Severn	Bristol University UK	Various projects related to structural design for earthquake including risk assessment + data management		
Entec Europe Ltd	Simon Montague	Newcastle upon Tyne UK	Recent activities include risk assessment of several major structures and buildings		
ETC (UK)	Phil O'Keefe	Rwanda	Humanitarian response	Evaluation.	1990 - 95
ETC (UK)	Phil O'Keefe	Somalia	Humanitarian response	Evaluation.	1992 - 95
Flood Hazard Research Centre	Colin Green	Europe	Euroflood 2	Development of a flood hazard management strategy for Europe.	1994 - 96
Food Studies Group	Graham Eele	Ethiopia	Establishment of a food security reserve		1 month
Food Studies Group	Graham Eele	Kenya	Recommendations for drought response		1 month
Food Studies Group	Graham Eele	Angloa	Plan of action for nutrition		2 months

Organisation and Contact	Location	Project	Objectives	Duration
Food Studies Group	Kurdish Northen Iraq	Vulnerability assessment		1 month
Food Studies Group	Southern Africa	Drought management training workshops		1 year
Geology Dept University of Bristol	Chile and Costa Rica	Volcanic hazards	Study of geological hazards on active volcanoes.	
Geology Dept University of Bristol		Generic (EC+NERC)	Basic physics of volcanic eruptions.	
Gifford and Partners	Cairo	Cairo Mosque repair after earthquake in 1993	To advise on structural stabilization and repair of historic monuments in Cairo.	
Hadley Centre Meteorological Office	Developing countries	Technical co-operation	Improvements of capabilities of other national Met services - including disaster preparedness.	Ongoing
Hadley Centre Meteorological Office	NE Brazil and Sahel	Seasonal rainfall forecasts	Advise on the likelihood of drought in the areas on a seasonal basis.	Ongoing
Hadley Centre Meteorological Office	Caribbean, Pacific, SW Indian Ocean	Tropical cyclone advisories	Advise requesting centres of tracks of TC's as predicted by global weather model.	Ongoing
Hadley Centre Meteorological Office	Africa, Europe, Asia	Pollution response	Regional Specialised Meteorological Organisations for International Environmental Emergencies.	Ongoing
Hadley Centre Meteorological Office	UK	Pollution response (UK)	Provide meteorological support to emergency authorities in pollution emergency, chemical or nuclear.	Ongoing
Hadley Centre Meteorological Office	UK	Storm surge warnings	Provide timely warnings to appropriate authorities (NRA & police in England & Wales) of storm surges.	Ongoing
Hadley Centre Meteorological Office	UK	Severe weather warnings	Provide timely warnings of severe weather to emergency organisations and the public.	Ongoing

Contacts: Graham Eele (Food Studies Group); Prof R Sparks (Geology Dept, University of Bristol); A Trickebank (Gifford and Partners); W H Lyne (Hadley Centre Meteorological Office).

Organisation and Contact		Location	Project	Objectives	Duration
Hazard and Risk Management Studies (HARMS)	Tom Horlick-Jones	UK	Application of problem structuring methods in emergency management	Development of risk analysis based on emergency planning techniques.	Ongoing
Hazard and Risk Management Studies (HARMS)	Tom Horlick-Jones	UK	Urban hazard project	Clarify nature of urban hazards in old megacities.	Aug 1993 - June 95
Health & Nutritional Status Advisory Unit (HANSA)	Prof Colin Mills	Georgia and Armenia	AICF Consultancy	Anthropometric data analysis nutritional consultancy.	
Health & Nutritional Status Advisory Unit (HANSA)	Prof Colin Mills	Armenia	WomanAid International	Ration composition analysis and diet planning.	
Health & Nutritional Status Advisory Unit (HANSA)	Prof Colin Mills	South Africa	Fortified foods	Analysis and redesign of emergency rations.	
HelpAge International	Chris Beer	Sudan	Various	Displaced people.	Since mid 1980s
HelpAge International	Chris Beer	Cambodia	Resettlement programme	Also respond to problems caused by ongoing insecurity as necessary.	1992
HelpAge International	Chris Beer	Mozambique	Resettlement programme	Assist older returnees to reintegrate into their communities.	2 years
Homeless International	Ruth McLeod	Tamil Nadu India	TAMIL NADU Network construction and networks	Network & information sharing. Developing appropriate construction against monsoon winds.	1993 - 98
Homeless International	Ruth McLeod	Andhra Pradesh India		Fire damage reconstruction of houses. Supply of emergency response equipment i.e. tents, cooking utensils. Design of houses to withstand fire and wind.	1995 - 6
Homeless International	Ruth McLeod	Andhra Pradesh India	Construction and training	Construct 50 "safe houses". Establish task force of people trained in disaster management in 7 mitigation techniques with response equipment.	1995 - 6

Organisation and Contact		Location	Project	Objectives	Duration
Homeless International	Ruth McLeod	Jamaica	CRDC Information and education	Enhance information flows on housing and disaster mitigation in Jamaica.	1995 - 2000
Independent Consultant	Jon Bennett	Oxford	NGO Co-ordination in humanitarian assistance	To evaluate and enhance the institutional capacity of national NGO co-ordination bodies.	2 years
Independent Consultant	Jon Bennett	Rwanda Survey	Japan International Co-operation Agency (JICA)	To gather and present a briefing on British NGO and ODA inputs into Rwanda as part of a JICA evaluation.	January 1995
Independent Consultant	Eric Alley	Geneva	MCDA in disaster relief	To increase the volume and efficiency of the international community contribution to humanitarian operations.	Ongoing
Independent Consultant	David Oakley	UNCHS (HABITAT)	Flood management handbook	Advisory manual for provincial and local government.	
Independent Consultant	Rob Stephenson	London and Europe	Commercial market research in disaster related publications		Jan 1995
Independent Consultant	Kevin McKemey	University of Reading and Nicaragua	The environmental refugee	Research into the behavior response of resettled refugees in environmentally sensitive areas.	1991 - 1995
Independent Consultant	James Atkinson	Tanzania	European Commission food security programme	Maize supply and price stabilization.	
Independent Consultant	James V Henry	Various	Human resources	Developer of handbook/manual for emergency preparedness and response procedures.	
Independent Consultant	James V Henry	UK	Burundi	UK manager of Emergency Relief Programme- seeds, tools household items and conflict reduction.	12 months
Independent Consultant	Helen Young	UK, Southern Sudan, Kenya	Review of the Sudan Emergency Operations Consortium Programme in Southern Sudan	Acted as the nutritionist in a team of 4 independent consultants undertaking a wide range review of the programme since 1991.	Aug 1994 - Feb 95
Institute of Child Health	David Morley	Desk review	Research	Child nutrition and care during emergencies.	

Organisation and Contact		Location	Project	Objectives	Duration
Institute of Child Health	David Morley	Bosnia	Research	Review of food security and nutrition problems.	3 months
Institute of Child Health	David Morley	Southern Sudan	Consultancy	Evaluation of US Government relief strategies.	3 months
Institute of Terrestrial Ecology	Lloyd Anderson	UK, Europe and former Soviet Union	Pollution episodes	Monitoring, impact assessment, advisory service to government departments.	
Institute of Terrestrial Ecology	Lloyd Anderson	Morocco and Nigeria	Ecological aspects of drought mitigation	To determine the key factors affecting the ability of ecosystems to recover following drought and management implications.	5 yrs
Intermediate Technology	John Twigg	7 South Asian countries	Duryog Nivaran: South Asian initiative on disaster management	Institutional strengthening, networking, comparative research on disaster prevention and management, disseminating information.	1994 - 98
Intermediate Technology	John Twigg	Sudan, Kenya, Zimbabwe	Food security projects	Promote community based projects to address good food security needs.	Ongoing
Intermediate Technology	John Twigg	Peru	Disaster preparedness, mitigation and prevention	Training, information sharing, disaster planning at community level.	1994 - 8
Intermediate Technology	Nick Hall	India and The Philippines	Research into indigenous knowledge of disaster preparedness	Investigate communities knowledge and coping strategies.	1995
Intermediate Technology	John Twigg	Central and South America	La Red (network for social studies in disaster prevention in Latin America)	Institutional strengthening, networking, comparative research on disaster prevention, management, disseminating information.	1993 - 7
International Extension College	John McCall	Khartoum and Gedaref, Sudan	Sudan Open Learning Unit (SOLU)	To provide education to those adults who have missed out on formal education as a result of conflict /civil war in the region.	1984 ongoing

Organisation and Contact		Location	Project	Objectives	Duration
International Institute for Environment and Development (IIED)	David Satterthwaite	Various cities	Environmental problems in third world cities	Understanding scale and nature of environmental problems and their health impacts.	
International Institute for Environment and Development (IIED)	David Satterthwaite	General	Vulnerability and poverty	Understanding and acting on distinction between two urban areas.	
International NGO Training and Research Centre (INTRAC)	Brian Pratt	Oxford and Ethiopia	The social position of children in complex emergencies		1994 - 5
Karuna Trust	Peter Joseph	Latur Urmarga and Maharashtra India	Earthquake rehabilitation project	Hostel accommodation and educational support for children affected by earthquake & provision of community activities space.	1.5 years
Living Earth	Gwen Vaughan	Cameroon-South and North West Province	40 local projects	Watershed protection to environmental health and the prevention of erosion.	1989 onwards
Living Earth	Gwen Vaughan	Venezuela	71 local projects	Protection of environment.	
Llewelyn Davies Planning	Jon Rowland	General	Vulnerability and poverty	Understanding and acting on distinction between two urban areas.	
Llewelyn Davies Planning	Jon Rowland	Case studies in various cities	Environmental problems in cities	Understanding scale and nature of environmental problems and their health impacts.	
Marie Stopes International	Patricia Hindmarsh	Former Yugoslavia	Reproductive health kits	To assist traumatized refugee women.	2 years
Marie Stopes International	Patricia Hindmarsh	Former Yugoslavia	Emergency humanitarian aid	To assist traumatized refugee women.	2 years
Medical Emergency Relief International (MERLIN)	Christopher Besse	SW Rwanda	IDP health care + rehabilitation	Medical supplies, restoration of health structures training.	

Organisation and Contact		Location	Project	Objectives	Duration
Medical Emergency Relief International (MERLIN)	Christopher Besse	Goma, Zaire	Refugee health care	Relief to orphanage for cholera/dysentery epidemic.	
Medical Emergency Relief International (MERLIN)	Christopher Besse	Afghanistan	Public health	Training in WHO protocols. Rehabilitation training.	
Medical Emergency Relief International (MERLIN)	Christopher Besse	Goma, Zaire	Refugee health care	Adult camps - all health care aspects.	
Natural Resources Institute	Jim Williams	Namibia, Jordan, Algeria, Malawi, Ethiopia	Direct satellite reception for local environmental monitoring and management.	Use of a low cost robust system for local satellite data reception and product generation.	
Natural Resources Institute	Jim Williams	Namibia, Nigeria, Sri Lanka, Zimbabwe, Paraguay	High resolution and remote sensing	Provides a comprehensive service for the interpretation and analysis of remote sensing.	
Natural Resources Institute	Jim Williams	Ethiopia, Zimbabwe, Mozambique, Zambia	Food security and famine preparedness	To develop procedures for effective and efficient distribution for food aid and strategies for maintaining food security in potential famine areas.	
Natural Resources Institute	Jim Williams	Brazil, Bangladesh, Ghana, Sri Lanka	Forestry	Forest inventory and management project.	
NERC Unit for Thematic Information Systems (NUTIS)	Prof Geoffrey Wadge	Geneva	Statistical modeling of volcanic hazards	Development of general GIS/statistics modeling.	1992 - 1995
NERC Unit for Thematic Information Systems (NUTIS)	Prof Geoffrey Wadge	Sicily and Costa Rica	Radar interferometry for volcano monitoring	Development of new remote sensing monitoring technique.	1995 - 1997
Noble Denton Weather Services Ltd	Howard Lawes	Japan	Research project to define accurately the regional exposure to typhoons in Japan (proposed)		

Organisation and Contact	Location	Project	Objectives	Duration
Overseas Development Administration, Emergency Aid Department (EMAD) Janet Douglas	Global	Funding for organisations to carry out various disaster mitigation, preparedness and response activities	SEE ODA CAPTION BOX IN FUNDING SECTION FOR A SUMMARY OF PROJECTS.	
Overseas Development Institute (ODI) John Borton	UK	Relief and Rehabilitation Network (RRN)	To provide information to NGO, UN, academic and bilateral agencies involved in relief.	1994 - 97
Overseas Development Institute (ODI) John Borton	UK	Research and consultancy	Trends in international aid policy/ responses to complex emergencies.	
Overseas Development Institute (ODI) John Borton	Rwanda	Multi-donor evaluations of international interventions		1995
Oxfam, Emergency Department Nick Stockton	South	Numerous emergency relief programme	Relief of distress and suffering.	1-2 years
Oxfam, Emergency Department Nick Stockton	Oxfam headquarters UK	Numerous emergency preparedness programmes	Preparedness to respond to emergencies.	
Oxford Centre for Disaster Studies (OCDS) Ian Davis, David Sanderson	Oxford, UK	Audit of UK Assets	To obtain a coherent overview of the current UK assets available in the field of disaster mitigation and preparedness.	Nov 1994 - May 95
Oxford Centre for Disaster Studies (OCDS) David Sanderson	Peru	Urban risk reduction	Reduction of risk to most vulnerable groups in Lima.	Oct 1995 - Apr 96
Oxford Centre for Disaster Studies (OCDS) Ian Davis, Roger Bellers	Turkey	Technical co-operation project with AFEM	Subject preparation of a manuals materials, training of trainers, database design.	April 1994 - April 96
Oxford Centre for Disaster Studies (OCDS) Roger Bellers	Philippines	Training in disaster management		Jan 1995 - Sept 95

Organisation and Contact		Location	Project	Objectives	Duration
Oxford Centre for Disaster Studies (OCDS)	David Sanderson	Zimbabwe, Mozambique, Rwanda	Training in disaster management in Africa	To increase the capacity of NGOs to develop, manage and sustain their own disaster mitigation and preparedness programmes.	June 1995, Sept 95, May 96
Post War Reconstruction & Development Unit (PRDU)	Sultan Barakat	Sri Lanka, South Africa, Macedonian, Mexico, Lebanon	Integrating cultural heritage into national disaster preparedness planning in 5 regional workshops		June 1995, July 95, Oct 95, Jan 96,
Responding To Conflict	Sue Williams	Africa, Pakistan, Afghanistan, Fiji, Former Yugoslavia	Various projects		
Rural Resources Management Ltd	Steve Jones	Bangladesh	Specialist advice to flood action plan		1990 -1995
Rural Resources Management Ltd	Steve Jones	Western India	Development of drought preparedness strategy for poor tribal communities		1995
Saferworld	Hugh Venables	Europe	Seminar series relating to the true cost of conflict and conflict prevention/management	To highlight the true cost of conflict and provide a forum for the discussion of priority areas for 1995.	Feb - May 1995
Saferworld	Hugh Venables	Bristol and London UK	Conflict management programme	To research and build support for proposal to enhance the international prevention and management of inter-state conflict.	Ongoing
Save the Children Fund UK	John Seaman	Rwanda, Zaire, Tanzania, Ethiopia, Sudan, Nepal		Range of social issues, nutrition, food, planning, health information activities.	
School of Oriental and African Studies (SOAS)	Prof J Allan	Sahelian Africa	Earth observations and geo-information for logistical applications in refugee relief	To develop information systems based in macro-level / remote sensing data, and ground level data.	
Scottish Catholic International Aid Fund (SCIAF)	Joan Mcerlean	Bosnia	Bosnia winter emergency programme	Essential goods provision-food, clothing etc.	Oct 1994 - Mar 95

Organisation and Contact		Location	Project	Objectives	Duration
Scottish Catholic International Aid Fund (SCIAF)	Joan Mcerlean	Rwanda	Relief/rehabilitation (various programmes)	Support of work of CARITAS, local NGOs for immediate and longer term needs of unaccompanied children etc.	Ongoing
Scottish Catholic International Aid Fund (SCIAF)	Joan Mcerlean	Malawi	Drought relief	Provision of food, seeds etc.	Sept 1994 - July 95
Scottish Catholic International Aid Fund (SCIAF)	Joan Mcerlean	Southern Sudan	SEOC rehabilitation and food security	Emergency food and water rehabilitation.	Jan - May 95
Scottish Catholic International Aid Fund (SCIAF)	Joan Mcerlean	Nuassa Mozambique	RRR programme	Rehabilitation(seeds/tools/education).	1994 - 95
Scottish Catholic International Aid Fund (SCIAF)	Joan Mcerlean	Manquete (Cunene)	Reintegration for returnees	Rehabilitation (seeds, tools, education).	1994 - 95
Sedgwick Global	Charles Toomer	EW Payne Subsidiary London, UK	Examination of computer/GIS based risk assessment models	Acquire an appropriate system.	
Sedgwick Global	Charles Toomer	Caribbean	Risk assessment - hurricanes	Establish "worst case" hurricane tracks for major clients location in Caribbean and adjust insurance buying accordingly.	Ongoing
Silsoe Research Institute	Derek Sutton	Bangladesh, Indonesia, Africa	Researching ergonomic problems of men and women in agriculture		
Silsoe Research Institute	Derek Sutton	Africa	Currently researching the problems of small farms on steep slopes in humid tropics and semi arid regions		
SOS Sahel	Duncan Fulton	Elain, Sudan	Natural forest management project		1994 - 1997
SOS Sahel	Duncan Fulton	Wollo Ethiopia	Agricultural support project	Income generation, livestock agricultural and water development.	1994 - 1997
SOS Sahel	Duncan Fulton	Kindo Kosha Ethiopia	Food security project	To improve long term food security.	

Organisation and Contact	Location	Project	Objectives	Duration
SOS Sahel Duncan Fulton	Sudan and Red Sea	Khor Arbaat rehabilitation project	Improving traditional water harvesting of flood waters.	
SOS Sahel Duncan Fulton	Bankass Mali	Bankass environmental project	Traditional conservation techniques against drought.	1992 - 1995
SOS Sahel Duncan Fulton	Touininvan Mali	Community environment project	Natural resource management.	1994 - 1997
South Bank University Tim Allen	Africa	Returning refugees, effects of war and human rights issues	Research, teaching and publications.	Since late 1980s
Systems Group, The Open University V Bignell	UK	Publication of disaster and failure case studies	Awareness.	
Tear Fund Mike Wall	Ngora Tanzania	Tanzania refugee project	Community services, health care and school construction for refugees.	Since April 1994
Tear Fund Mike Wall	Goma Zaire	Zaire refugees project	Community services and camp management to Rwanda refugees.	Since August 1994
The Centre for Crisis Psychology PE Hodgkinson	UK	Medical-legal response	Provision of assessment of extent of trauma effect.	Daily
The Research Centre Sue Ellis	Croatia	Analysis & evaluation of shelter projects and policies for refugees in Croatia	Development of guidelines for improved shelter response for refugees.	2 years
The Society for Earthquake and Civil Engineering Dynamics (SECED) Mary Kensella	UK	Needs assessment service for disaster situations	To provide assessment of 'engineering' needs.	
The Society for Earthquake and Civil Engineering Dynamics (SECED) Mary Kensella	Computer Based UK	Maintenance & operation of register of disaster relief personnel	Core activity and recruitment.	Ongoing
Trocaire Niall Toibin	Gedo Region Somalia		Water and health.	
Trocaire Niall Toibin	Eastern Equatorial and Southern Sudan		Medical.	

Organisation and Contact	Location	Project	Objectives	Duration
Trocaire — Niall Tcibin	South West Rwanda	Rwanda emergencies project	Medical relief and agricultural services and tools.	8 months
UK National Coordination Committee for the IDNDR — Jacqueline Baines	London, Cardiff, Edinburgh, Belfast	IDNDR day	To plan a series of lectures on disaster mitigation and preparedness.	1 day
UK-Med — A Redmond	Global	Disaster assessment capability (UNDAC Team)	To provide an assessor at 3 hours notice to move in the event of a disaster.	Ongoing
UK-Med — A Redmond	Sarajevo and UK	Ophthalmology programme	To establish and support an effective ophthalmology service in Bosnia; cataract treatment programme; artificial eye service.	3 years from April 1995
UK-Med — A Redmond	Sarajevo	Operation Phoenix	Support, hospital service, essential drugs/equipment and emergency department.	9 months from March 1995
UK-Med — A Redmond	Society of Apothecaries London UK	Diploma in Medical Care For Catastrophes	Provide 2 examiners and courses for the Diploma.	Ongoing
United Kingdom Jewish Aid — Mr Harris	Former Yugoslavia and Israel	Helping the Helpers	Training professionals to cope with trauma/refugees in former Yugoslavia.	
Water Engineering and Development Centre (WEDC) — Bob Reed	WEDC Loughborough UK	Research	Alternative approaches to the provision of water and sanitation advising field workers when required.	
Water Engineering and Development Centre (WEDC) — Bob Reed	WEDC Loughborough UK	Training	Training more specialists in the basic of emergency water supply and sanitation.	2 weeks
World Vision UK — Susan Barber	Cambodia	Relief and Rehabilitation	Relief/rehabilitation for internally displaced persons, shelter, infrastructure, food security.	Ongoing since 1990
World Vision UK — Susan Barber	Rwanda	Rwanda programme	Support rehabilitation in Rwanda particularly related to agriculture and children.	

The directory

Directory of organisations
IDNDR focal points
Directory of individuals

Directory of organisations

The following directory of UK-based organisations represents the information returned in the questionnaires. Any spaces indicates areas unanswered by respondents.

Action Health

The Gate House 25 Gwydir Street
Cambridge CB1 2LG

Tel   01223 460853
Fax   01223 460853
E Mail

Contact   Kate Graham
            Director

Date Formed 1984   No. staff   4
Income   £117 992 Expenditure   £148 226

Mission Statement

To work in partnership with local communities to identify health needs and develop appropriate health programmes by the transfer of skills, with an emphasis on utilisation of local resources.

Action Water Charity

Mount Hawke Truro
Cornwall TR4 8BZ

Tel   01209 715385
Fax   01209 715385
E Mail

Contact   Neil Battersby
            Founder Trustee

Date Formed 1985   No. staff   1 full time
Income   £25,000 Expenditure   £20,000

Mission Statement

Action Water is a unique charity that refurbishes water equipment for further use in community projects in the under-developed world. Items such as water drilling rigs and water tankers are overhauled and modified for use in hotter climates.

ActionAid

Hamlyn House MacDonald Road Archway
London N19 5PG

Tel   0171 282 4202
Fax   0171 281 2076
E Mail

Contact   Margie Buchanan-Smith
            Head of Emergency Unit

Date Formed 1972   No. staff   200 UK, 2500 o/s
Income   Expenditure   £6.7 million

Mission Statement

To work with some of the world's poorest children, families and communities to enable them to alleviate their poverty, and secure lasting improvements in the quality of their lives.

Adventist Development and Relief Agency (ADRA)

119 St Peter's Street
St Albans Herts AL1 3EY

Tel   01727 860331
Fax   01727 866312
E Mail

Contact   John Arthur
            Executive Director

Date Formed 1977   No. staff   4 UK, 3000 o/s
Income   £120 million Expenditure

Mission Statement

To provide assistance in situations of crisis or chronic distress and work towards the development of long-term solutions with those affected. Also to actively support communities in need through a portfolio of development activities which are planned and implemented co-operatively.

Afghanaid

292 Pentonville Road
London N1 9NR

Tel   0171 278 2832
Fax   0171 837 8155
E Mail

Contact   Wynn Flaten
            Managing Director

Date Formed 1983   No. staff   130
Income   £1 million Expenditure   £1 million

Mission Statement

To provide humanitarian assistance to the people of Afghanistan, through emergency relief, rehabilitation and development.

Aids, Care, Education and Training (ACET)

P.O. Box 3693
London SW15 2BQ

Tel 0181 780 0400
Fax 0181 780 0450
E Mail

Contact Maurice Adams

Date Formed 1988 No. staff 50
Income £1.2 million Expenditure £1.2 million

Mission Statement

In communities around the world ACET provides unconditional care for those with AIDS, practical education and training about AIDS and its prevention.

Agency for Research and Co-operation in Development (ACORD)

Francis House Francis Street
London SW1P 1DQ

Tel 0171 828 7611
Fax 0171 976 6113
E Mail acord @ gn.apc.org

Contact Peter James
Programme Director

Date Formed 1976 No. staff 29 UK 5 o/s
Income £ 7.1 million Expenditure

Mission Statement

ACORD is a broad-based international corporation of NGOs. The corporation's main role is to help strengthen or establish local NGO structures to promote self-reliant participatory development.

Agricultural Extension and Rural Development Department (AERDD)

The University of Reading 3 Earley Gate P. O. Box 238
The University Reading RG6 6AL

Tel 01734 318119
Fax 01734 261244
E Mail easgarfo@reading.ac.uk

Contact Dr Chris Garforth
Head of Department

Date Formed 1965 No. staff 20
Income £1 million Expenditure

Mission Statement

Teaching and research related to communication, management and social aspects of rural resources- based development.

Alexander Howden Group Limited

8 Deveonshire Square
London EC2M 4PL

Tel 0171 623 5500
Fax 0171 621 1511
E Mail

Contact

Date Formed No. staff
Income Expenditure

Mission Statement

Appropriate Health Resources and Technology Action Group (AHRTAG)

Farringdon Point 29-35 Farringdon Road
London EC1M 3JB

Tel 0171 242 0606
Fax 0171 242 0041
E Mail ahrtag@gn.apc.org/
ahrtag@geo2:geonet.de

Contact

Date Formed 1977 No. staff 27
Income £1,45 million Expenditure £1,48 million

Mission Statement

The appropriate Health Resources and Technologies Action Group (AHRTAG) is an international development agency established in 1977 to strengthen the management and practice of primary health care to community based rehabilitation in developing countries by maximising the use of information.

Book Aid International

39-41 Coldharbour Lane Camberwell
London SE5 9NR

Tel 0171 733 3577
Fax 0171 978 8006
E Mail RLS@gn.apc.org

Contact David Membrey
Deputy Director

Date Formed 1954 No. staff 20 full time
Income £700 000 Expenditure £700 000

Mission Statement

Book Aid International works in partnership with organisations in developing countries to support their work in literacy, education, training and publishing by providing books, journals, other printed materials and alternative media, which helps give people the chance to realise their potential and contribute to the development of their societies.

British Association for Immediate Care

Bay 6-7 Black Horse Lane
Ipswich Suffolk IP1 2EF

Tel   01473 218407
Fax   01473 280585
E Mail

Contact   Ron Bailey
              Chief Executive

Date Formed 1977   No. staff   2
Income   Expenditure

Mission Statement

Provision of medical care at emergencies. The aims of BASICS are to foster co-operation between existing immediate care schemes and to encourage the formation and extension of schemes in the United Kingdom; to develop and strengthen co-operation between all services in dealing with emergencies; to encourage and assist research into all aspects of immediate care and accident prevention; to raise the standards of immediate care and training; to produce publications for the dissemination of information, and to evaluate specialist emergency equipment.

British Consultants Bureau

1 Westminster Palace Gardens 1-7 Artillery Row
London SW1P 1RJ

Tel   0171 222 3651
Fax   0171 2223644
E Mail   ian@bcbtrack.demon.co.uk.

Contact   T A Boam
              Director

Date Formed 1965   No. staff 10
Income   Expenditure

Mission Statement

Promote British consultancy of all disciplines world-wide, provide a forum in which consultants can meet, provide information on public procurement of services and act as a focal point for governments.

British Geological Survey

Kingsley Dunham Centre Keyworth Nottingham NG12 5GG

Tel   0115 9363100
Fax   0115 9363165
E Mail

Contact   Paul Gostelow

Date Formed   No. staff 800
Income   Expenditure

Mission Statement

Geological surveys of UK landmark research (pure and applied) to support International/national wealth creation and environmental protection.

British Red Cross Society Headquarters

9 Grosvenor Crescent London SW1X 7EJ

Tel   0171 235 5454
Fax   0171 235 0397
E Mail

Contact   Mike Adamson
              Head International Development

Date Formed 1870   No. staff   289
Income   £34 million Expenditure   £33 million

Mission Statement

To furnish aid to the sick and wounded in time of war. The improvement of health, the prevention of disease and the mitigation of suffering throughout the world.

Building and Social Housing Foundation

Memorial Square Coalville
Leicestershire LE67 3TU

Tel   01530 510444
Fax   01530 510332
E Mail

Contact   Diane Daicon
              Research Officer

Date Formed 1976   No. staff 5
Income   Expenditure

Mission Statement

Research into the development, management and construction of residential housing.

Cambridge Architectural Research Ltd

The Eden Centre 47 City Road Cambridge CB1 1 DP

Tel   01223 460475
Fax   01223 464142
E Mail

Contact   Antonios Pomonis
              Director

Date Formed 1986   No. staff 10
Income   Expenditure

Mission Statement

CAR is an independent consultancy with 15 years experience in disaster mitigation, preparedness and protection. It provides specialist advice in vulnerability and risk assessment, hazard mitigation, planning and geographical information systems.

Care International (UK)

36 - 38 Southampton Street London WC2E 7AF

Tel 0171 379 5247
Fax 0171 379 0543
E Mail

Contact James Fennell
Head of Emergency Operations

Date Formed 1985 No. staff 40
Income Expenditure

Mission Statement

To combat poverty through relief during emergencies.

Cargil Attwood Consultants

8 Teddington Park Teddington TW11 8DA

Tel 0181 977 8091
Fax 0181 9431393
E Mail

Contact Tom Attwood
Chief Executive

Date Formed 1965 No. staff 25
Income Expenditure

Mission Statement

International management and training consultants.

Catastrophe Reinsurance

DYP Group Limited Bridge House 181 Queen Victoria Street London EC4V 4DD

Tel 0171 236 3223
Fax 0171 489 1487
E Mail

Contact Graham Village
Editor

Date Formed 1993 No. staff
Income Expenditure

Mission Statement

Catholic Fund for Overseas Development (CAFOD)

2 Romero Close Stockwell Road
London SW9 TYUK

Tel 0171 733 7900
Fax 0171 274 9630
E Mail cafod@gn.apc.org

Contact Tony Hardiment
Head, Communication Section

Date Formed 1962 No. staff 120
Income £25 million Expenditure

Mission Statement

CAFOD's concern is for all that reates to human development and to the plight of the poor in a world divided between a rich North and a poor South

Centre for African Studies

University of Cambridge Free School Lane Cambridge CB2 3RQ

Tel 01223 334396/9
Fax 01223 334396
E Mail african-studies@lists.cam.a

Contact Paula Munro
Administrator

Date Formed 1965 No. staff 3
Income Expenditure

Mission Statement

Centre for Arid Zone Studies

Thoday Building University of Wales Bangor Gwynedd LL57 2UW

Tel 01248 382346
Fax 01248 364 717
E Mail

Contact D Wright

Date Formed 1984 No. staff 30
Income £980 000 Expenditure £960 000

Mission Statement

Application of agricultural forestry and environmental sciences to development in arid, semi-arid and seasonally arid areas.

Centre for Developing Areas Research (CEDAR)

Department of Geography Royal Holloway University of London Egham London TW20 0EX

Tel 01784 443651
Fax 01784 472836
E Mail D. Simon @ rhbnc.ac.uk

Contact Dr David Simon
Director

Date Formed 1988 No. staff 11
Income Expenditure

Mission Statement
Research, consultancy and teaching on geography and development in the Third World.

Centre for Development and Emergency Planning (CENDEP)

Oxford Brookes University Gipsy Lane
Headington Oxford OX3 0BP

Tel   01865 483413
Fax   01865 483298
E Mail

Contact   Hugo Slim
           Co-Director

Date Formed 1994   No. staff 3
Income   Expenditure

Mission Statement

To improve understanding of people's experience of complex emergencies and the impact of humanitarian assistance and peacekeeping programmes in order to inform current policy and practice.

Centre for Environmental and Human Settlements (CEHS)

School of Planning and Housing
Edinburgh College of Art Heriot-Watt University Lauriston Place Edinburgh Scotland EH3 9DF

Tel   0131 2216000
Fax   0131 2216163
E Mail   tcpsm@uk.ac.hw.vaxa

Contact   John B Leonard, Director

Date Formed 1991   No. staff 2
Income   Expenditure

Mission Statement

Postgraduate teaching and research facilities for human settlements development and environment.

Centre for International Health

Room 12 Upper Ground Floor University of Wales College of Medicine Heath Park Cardiff CF4 4XN

Tel   01222 742323
Fax   01222 742898
E Mail

Contact   Eileen Darby

Date Formed 1984   No. staff 4
Income   Expenditure

Mission Statement

To help improve the health of vulnerable groups within developing country populations by educating health professionals, undertaking research, and contributing to debate in priority areas. The CIH will adopt a multi-disciplinary approach to issues in public health policy, health services delivery, and disaster preparedness with specific reference to the needs of women, children, the elderly, disabled, the displaced and the poor. It will also address the needs of persons resident or refugees from developing countries in the UK.

Centre for the Sciences of Food and Nutrition (CSFN)

Oxford Brookes University Gipsy Lane Headington Oxford. OX3 OBP.

Tel   01865 483818
Fax   01865 484017
E Mail

Contact   Dr J Henry
          Director

Date Formed 1988   No. staff 7
Income   Expenditure

Mission Statement

CSFN is an interdiscipliniary unit that aims to integrate the needs of the food industry, agriculture, retailing and nutrition.

Centre for the Study of African Economies

21 Winchester Road Oxford OX2 6NA

Tel   01865 274557
Fax   01865 274558
E Mail   ECONPGB@vax.ox.ac.uk

Contact   Philippa Bevan
          Research Officer

Date Formed 1991   No. staff
Income   Expenditure

Mission Statement

Centre for Urban & Regional Studies

School of Public Policy JG Smith Building
University of Birmingham Edgbaston
Birmingham B15 2TT

Tel   0121 4145021
Fax   0121 4143279
E Mail   M.Duffield@bham.ac.uk

Contact   Mark Duffield

Date Formed 1985   No. staff 70
Income   Expenditure

Mission Statement

Excellence and innovation in public policy

Charity Projects (Comic Relief)

1st Floor 74 New Oxford Street
London WC1A 1EF

Tel 0171 436 1122
Fax 0171 436 1541
E Mail

Contact Maggie Baxter
Grants Director

Date Formed 1985 *No. staff* 28
Income £18.4 million *Expenditure* £18.4 million

Mission Statement

Charity Projects exists to help disadvantaged people in the UK and in Africa realise their aspirations and potential. We do this by raising new money from the public, inform and educating, allocating the funds we get and learning from experience.

Christian Aid

P. O. Box 100
London SE1 7 RT

Tel 0171 620 4444
Fax 0171 620 0719
E Mail geonet geo2:caid/greennet g

Contact Jenny Borden
Overseas Director

Date Formed 1945 *No. staff* 220
Income £42 million *Expenditure* £42 million

Mission Statement

To strengthen the poor through relief, rehabilitation, development, education in the North, advocacy and campaigning.

Climatic Research Unit

School of Environmental Science University of East Anglia
Norwich NR4 7TJ

Tel 01603 592088
Fax 01603 507784
E Mail M.Hulme@uea.ac.uk

Contact Mike Hulme
Senior Research Associate

Date Formed *No. staff* 20
Income *Expenditure*

Mission Statement

Concern Worldwide

248-250 Lavender Hill
London SW11 1LJ

Tel 0171 738 1033
Fax 0171 738 1032
E Mail

Contact Nick Guttmann
Logistics Advisor (London)

Date Formed 1968 *No. staff*
Income *Expenditure*

Mission Statement

Concern is devoted to the relief, assistance and advancement of peoples in need in less developed areas of the world.

Cranfield Disaster Preparedness Centre (CDPC)

Cranfield University RMCS Shrivenham
Swindon Wilts SN6 8LA

Tel 01793 785287
Fax 01793 782179
E Mail

Contact Ken Westgate
Director

Date Formed 1985 *No. staff* 6
Income £309 000 *Expenditure* £336 000

Mission Statement

To be a leader in the promotion of development-based disaster management in the developing nations, particularly those in Africa.

Cranfield School of Management

Cranfield University Bedford MK43 0AL

Tel 01234 751122
Fax 01234 751806
E Mail

Contact John Hailey

Date Formed *No. staff* 250
Income £15 million *Expenditure*

Mission Statement

To build the management capacity of NGOs and development agencies by undertaking research, provide training to develop the management capacity of NGO managers.

Crown Agents

St Nicholas House St Nicholas Road
Sutton Surrey SM1 1EL

Tel	0181 6433311
Fax	0181 6438232
E Mail	

Contact   Alan Matthews
               Divisional Director, Emergencies

Date Formed	1950s	No. staff	700
Income	£42.5 million	Expenditure	£37.3 million

Mission Statement

Our aim is to be the foremost international supplier of goods and services, and through our efficiency and integrity, to enable nations to make the very best of available resources.

Datum International

101 High Street Marshfield Chippenham
Wiltshire SN6 8LA

Tel	01225 891 426
Fax	01225 892 092
E Mail	datum@gn.apc.org

Contact   James Lewis
               Director

Date Formed	1980	No. staff	
Income		Expenditure	

Mission Statement

Development planning for vulnerability reduction and survival: with the integration of appropriate building design and construction, taking account of the potential and of the aftermath of ecological change, natural disasters, conflict and civil strife.

Department of Civil Engineering, Imperial College

South Kensington
London SW7 2BU

Tel	0171 589 5111
Fax	0171 823 8525
E Mail	c.j.kerr@ic.ac.uk

Contact   C J Kerr
               Department Administrator

Date Formed	1907	No. staff	110
Income	£4 million	Expenditure	£4 million

Mission Statement

Research and teaching in all aspects of civil, structural and environmental engineering.

Department of Geography, University of Liverpool

Liverpool L69 3BX

Tel	0151 794 2876
Fax	0151 794 2866
E Mail	J654@liv.ac.uk

Contact   David Chester

Date Formed	1978	No. staff	1
Income		Expenditure	

Mission Statement

Department of Civil Engineering, University of Portsmouth

Portsmouth PO1 3QL

Tel	01705 842423
Fax	01705 842521
E Mail	blee@civl.port.ac.uk

Contact   Brian Lee
               Head of Civil Engineering

Date Formed	1971	No. staff	
Income		Expenditure	

Mission Statement

To apply an understanding of applied aerodynamics and fluid mechanics to the solution of disaster related problems concerning wind flow.

Department of Civil Engineering, University of Nottingham

University Park
Nottingham NG7 2RD

Tel	0115 9515151
Fax	0115 9513898
E Mail	

Contact   Reader in Fluid Engineering

Date Formed		No. staff	
Income		Expenditure	

Mission Statement

Department of Geography, University of Cambridge

Downing Place
Cambridge CB2 3EN

Tel 01223 333 399
Fax 01223 333 392
E Mail clive@lithos.jpl.nasa.gov

Contact Clive Oppenhiemer

Date Formed 1988 No. staff
Income Expenditure

Mission Statement

Research and teaching

Department of Social Anthropology, University of Manchester

Roscoe Building 5th Floor Brunswick Street
Manchester M13 9PL

Tel 0161 275 3999
Fax 0161 2753970
E Mail D.A.Turton@man.ac.uk

Contact David Turton
Editor of Disasters

Date Formed 1973 No. staff
Income Expenditure

Mission Statement

Department of the Environment

Room A4.32 Romney House 43 Marsham St
London SW1P 3PY

Tel 0171 276 8687
Fax 0171 276 8639
E Mail

Contact Colin Wright
Water Resources + Reservoirs

Date Formed No. staff 4800
Income £46 million Expenditure

Mission Statement

Environment protection

Department of Community Medicine

Fenners Gresham Road Cambridge CB1 2ES

Tel 01223 336590
Fax 01223 336584
E Mail

Contact Peter Baxter

Date Formed 1980 No. staff 13
Income Expenditure

Mission Statement

Public health.

Department of Geography, Chester College

Cheyney Road Chester GH1 4BJ

Tel 01244 375444
Fax 01244 373379
E Mail

Contact Martin Degg

Date Formed 1985 No. staff 3
Income Expenditure

Mission Statement

Research and teaching in the field of natural hazards, vulnerability disasters and development.

Department of Geography, London School of Economics and Political Science

Houghton Street London WC2A 2AE

Tel 0171 955 7577
Fax 0171 955 7721
E Mail

Contact Tom Horlick-Jones
Date Formed 1984 No. staff 2 Part-time
Income Expenditure

Mission Statement

Department of Geography, University of Durham

South Road Durham DH1 1QT

Tel 0191 374 2448
Fax 0191 374 2456
E Mail

Contact Ewan Anderson
Reader in Geopolitics

Date Formed 1986 No. staff 1
Income Expenditure

Mission Statement

To develop a global database, to develop an early warning system, and to identify indicators for communities.

Development Planning Unit, University of London

9 Endsleigh Gardens London WC1H OED

Tel 0171 388 7581
Fax 0171 387 4541
E Mail dpu@ucl.ac.uk

Contact Babar Mumtaz

Date Formed 1976 No. staff 15
Income £1 million Expenditure £1 million

Mission Statement

To develop institutional capacity of developing countries especially those institutions and individuals dealing with urban development and housing.

Devon Aid

Lower Beer Uplowman Tiverton Devon EX16 7PF

Tel 01884 821239
Fax 01884 821239
E Mail

Contact Robert L P Hodgson

Date Formed 1980 No. staff
Income Expenditure

Mission Statement

Technical support for disaster relief and mitigation.

Durham University Business School

The University of Durham Mill Hill Lane Durham DH1 3LB

Tel 0191 374 7326
Fax 0191 374 7163
E Mail

Contact

Date Formed 1981 No. staff
Income Expenditure

Mission Statement

Education and research.

Earth Observation Sciences Ltd

Broadmede Farnham Business Park
Farnham Surrey GU9 8QL

Tel 01252 721444
Fax 01252 716168
E Mail DaveS@eos.co.uk

Contact D R Sloggett
Business Development Direct

Date Formed 1986/1991 No. staff 90
Income £3.4 million Expenditure

Mission Statement

To develop and apply systems expertise to environmental monitoring.

Earth Resources Centre

University of Exeter Laver Building North Park Road Exeter EX4 4QE

Tel 01392 263909
Fax 01392 263907
E Mail

Contact John R Merefield
Research Manager

Date Formed 1990 No. staff 20
Income £250 000 Expenditure £200 000

Mission Statement

To apply scientific solutions to environmental problems.

Earthquake Engineering Research Centre

Department of Civil Engineering Queens Building University Walk Bristol University Bristol BS8 1TR

Tel 01272 287 706
Fax 01272 287 783
E Mail

Contact R T Severn
Director of Earthquake Engineering

Date Formed 1985 No. staff 15
Income £350 000 Expenditure £350 000

Mission Statement

To mitigate the effects of earthquake on man made systems.

Earthscan Publications Ltd

120 Pentonville Road London N1 9JN

Tel 0171 278 0433
Fax 0171 278 1142
E Mail

Contact

Date Formed No. staff
Income Expenditure

Mission Statement

ECHO International Health Service

Ullswater Crescent Coulsdon
Surrey CR5 2HR

Tel 0181 6602220
Fax 0181 668 0751
E Mail

Contact Keith Slatter
Chief Executive

Date Formed 1972 No. staff 60
Income £7.5 million Expenditure £7.5 million

Mission Statement

ECHO is a health orientated supply charity serving poor/disadvantaged communities worldwide, principally through the supply of medicines, reduced equipment /consumable and providing related advisory services.

Entec Europe Ltd

Northumbria House Regent Centre Gosforth
Newcastle-upon-Tyne NE3 3PX

Tel 0191 213 5126
Fax 0191 2854631
E Mail

Contact Alan Gray
Marketing Director

Date Formed No. staff 1000
Income £57 million Expenditure

Mission Statement

Business consultancy with an environmental focus. Our consultants investigated the Piper Alpha oil platform disaster, the Kings Cross Fire (London Underground) and the Clapham Rail Disaster.

Environmental Change Unit

1A Mansfield Road Oxford OX1 3TB

Tel 01865 281180
Fax 01865 281181
E Mail tom.downing@ecu.ox.uk.ac

Contact Tom Downing

Date Formed 1991 No. staff 30
Income £250,000 Expenditure £250,000

Mission Statement

The climate impact and responses programme seeks to understand the short and long-term effects of climate variations on resources, economies and societies. Research on vulnerability assessment strives to improve famine early warning systems and mitigation of natural hazards.

ETC (UK)

117 Norfolk Street North Shields
Tyne and Wear NE30 1NQ

Tel 0191 296 1681
Fax 0191 296 1682
E Mail

Contact Phil O'Keefe

Date Formed 1972 No. staff 6
Income £400,000 Expenditure £400,000

Mission Statement

Participatory development in a sustainable manner.

Farm Africa

9-10 Southampton Place Bloomsbury
London WC1A 2DA

Tel 0171 430 0440
Fax 0171 430 0460
E Mail farmafricauk@gn.apc.org

Contact David Campbell
Executive Director

Date Formed 1985 No staff 10 UK, 100 o/s
Income £1 441 709 Expenditure £1 436 820

Mission Statement

FARM-Africa is committed to helping small farmers and herders of Africa to help themselves. We believe that these are the people who can contribute most towards breaking the cycle of famine and bring hope and new prosperity to neglected marginal communities.

Flint and Neill Partnership

14 Hobbart Place London SW1 W0HH

Tel 0171 234 9911
Fax 0171 235 1259
E Mail

Contact Mr Brian Smith
Senior partner

Date Formed No. staff
Income Expenditure

Mission Statement

Flood Hazard Research Centre

Middlesex University Queensway Enfield Middx EN3 4SF

Tel	0181 3625359
Fax	0181 3625403
E Mail	FHRC1@uk.ac.mdx.cluster
Contact	Maureen Fordham Research Centre Manager
Date Formed	1970
No. staff	15
Income	£300 000
Expenditure	£300 000

Mission Statement

Applicable research in the field of water and disaster.

Food Studies Group

Queen Elizabeth House University of Oxford 21 St Giles Oxford OX1 3LA

Tel	01865 270261
Fax	01865 514 468
E Mail	FSG@qeh.ox.ac.uk
Contact	Roger Hay Director
Date Formed	1979
No. staff	17
Income	£950 000
Expenditure	£930 000

Mission Statement

To support and strengthen the capacity of indigenous organisations to analyse and implement policy to promote economic growth, alleviate poverty and protect people from the impact of disasters.

Geology Department, University of Bristol

Wills Memorial Building Bristol BS8 1RJ

Tel	01272 287789
Fax	01272 253385
E Mail	steve.sparks@Bristol.ac.uk.
Contact	R S J Sparks Professor of Geology
Date Formed	1976
No. staff	2
Income	£150 000
Expenditure	

Mission Statement

Research on volcanic processes.

Gifford and Partners

Carlton House Ringwood Road Woodlands Southampton SO40 7HT

Tel	01703 813461
Fax	01703 813462
E Mail	
Contact	Alan Tricklebank Director
Date Formed	1954
No. staff	20
Income	£ 7 million
Expenditure	

Mission Statement

To provide engineering services (advice/design etc.) civil, structural, environmental to clients that are appropriate, proportional, excellent and provide good value for money.

Global Seismology Research Group

British Geological Survey Murchison House West Mains Road Edinburgh EH9 3LA

Tel	0131 667 1000
Fax	0131 667 1877
E Mail	cwab@va.nmh.ac.uk
Contact	Dr Chris Browitt
Date Formed	1963
No. staff	30
Income	£ 1.5 million
Expenditure	£1.5 million

Mission Statement

To understand earthquake mechanisms, preparation processes and hazards as primary input to the planing and engineering problems of reducing risks.

Green Cross UK

Mayfied Centre 94 West Hill London SW15 2UH

Tel	0181 870 1843
Fax	0181 875 9606
E Mail	
Contact	Anne Grant Co-ordinator
Date Formed	1994
No. staff	1
Income	
Expenditure	

Mission Statement

To facilitate and help build capacity to prevent and respond to environmental disasters, and to foster global value shift with regard to the environment.

Hadley Centre Meteorological Office

London Road Bracknell
Berks RG12 2SZ

Tel   01344 420242
Fax   01344 854412
E Mail   whlyne @ meto.govt.uk

Contact   W H Lyne
          Manager Public Met Service

Date Formed 1854   No. staff   2 500
Income   £54.9 million   Expenditure   £97.5 million

Mission Statement

To excel in providing meteorological services that satisfy our customers current and future requirements.

Harvest Help

3-4 Old Bakery Row Wellington Telford TF1 1PS

Tel   01952 260699
Fax   01952 247158
E Mail

Contact   David White
          Director

Date Formed 1985   No. staff   5
Income   £400 000   Expenditure   £370 000

Mission Statement

Supporting long term development work for disadvantaged rural communities in Zambia and elsewhere in central Southern Africa.

Hazard and Risk Management Studies (HARMS)

c/o Dept. Geography London School of Economics and Political Science Houghton Street London WC2A 2AE

Tel   0171 955 7577
Fax   0171 955 7721
E Mail

Contact   Tom Horlick-Jones
          Acting Research Manager

Date Formed 1993   No. staff   1 full time
Income   £250 000   Expenditure

Mission Statement

Application of social science perspectives in hazard and risk management.

Health & Nutritional Status Advisory Unit (HANSA)

Rowett Research Institute Greenburn Road Bucksburn Aberdeen Scotland AB2 9SB

Tel   01224 716435
Fax   01224 716439 /716687
E Mail   hansa@rri.sari.ac.uk

Contact   C F Mills

Date Formed   No. staff   4 voluntary
Income   £20 574   Expenditure   £46 798

Mission Statement

Statistical analyses of anthropometric data to detect communities meriting nutritional priority treatment. Estimate nutrient input, balance and relationship to estimated need of communities at risks. Monitor effectiveness of nutritional aid from blood content of markers of nutrient status and/or anthropometric changes. Provide technical guidance on nutrient content of alternative foods, on nutrient stability and on nutritional factors.

Health Projects Abroad

P.O. Box 24 Bakewell
Derbyshire DE45 1ZW

Tel   01629 640051
Fax   01629 640054
E Mail

Contact   Simon Headington
          Director

Date Formed 1990   No. staff   2
Income   £220 000   Expenditure   £220 000

Mission Statement
Protect and preserve good health in developing countries.

HelpAge International

St James's Walk
London EC1R 0BE

Tel   0171 250 4411
Fax   0171 253 4814
E Mail   Greenet:Helpage@gn.apc.org

Contact

Date Formed 1970   No. staff   208
Income   £1.8 million Expenditure   £1.7 million

Mission Statement

Homeless International

Guildford House, 20 Queens Road Coventry CV1 3EG

Tel	01203-632802
Fax	01203-632911
E Mail	

Contact   Ruth McLeod
             Chief Executive

Date Formed	1987 as IYSH	No. staff	11
Income	£629 000	Expenditure	£502 000

Mission Statement

Homeless International is a registered charity based in the UK housing and construction sector. It has two main objectives: to provide financial and technical support to charitable organisations and groups working to improve shelter conditions of the poor in Asia, Africa, Latin America, the Caribbean and Europe; and to support the international exchange of information and experience on homelessness (supports long term disaster mitigation activities related to shelter & ensuring safety of housing in the event of natural disaster).

Institute for Development Policy and Management (IDPM)

University of Manchester Precinct Centre Oxford Rd
Manchester M13 9GH

Tel	0161 275 2800/2804
Fax	0161 273 8829
E Mail	IDPM@MAN.AC.UK

Contact   Joe Mullen

Date Formed		No. staff	
Income		Expenditure	

Mission Statement

To promote social and economic development, particularly within lower income countries and for disadvantaged groups, by enhancing the capabilities of individuals and organisations through education, training, consultancy, research and policy analysis.

Institute of Child Health

University of London 30 Guildford Street
London WC1N 1EH

Tel	0171 242 9789
Fax	0171 404 2062
E Mail	cich@ich.bpmf.ac.uk

Contact   Richard Longhurst
             Honorary Senior Lecturer

Date Formed	1970	No. staff	20
Income	£0.9 million	Expenditure	£0.9 million

Mission Statement

Working to promote better health, nutrition and welfare of children and families in less developed countries through excellence in teaching, research, consultancy and advocacy.

Institute of Civil Defence and Disaster Studies

Bell Court House 11 Bloomfield Street
London EC2M 7AY

Tel	0171 588 3700
Fax	0171 587 6350
E Mail	

Contact   J McK. Holloway
             Chairman General Council

Date Formed	1938	No. staff	
Income	£14 078	Expenditure	£11 696

Mission Statement

To advance and promote the national and international development of disaster management. To improve counter-disaster planning through studies into disaster situations which may recur and seeking logistical, technical, environmental solutions to problems. To suggest and develop specific technical solutions to imperative disaster problems associated with suffering and deprivation.

Institute of Development Studies (IDS)

University of Sussex Falmer
Brighton BN1 9RE

Tel	01273 606261
Fax	01273 621202
E Mail	s.j. maxwell@sussex.ac

Contact   John Toye
             Director

Date Formed	1966	No. staff	40
Income	£560 000	Expenditure	

Mission Statement

The IDS is a national research and training centre focusing on poverty in developing countries.

Institute of Risk Management

Lloyds Avenue House 6 Lloyds Avenue
London EC3N 3AX

Tel 0171 709 9808
Fax 0171 709 0716
E Mail

Contact Maureen B Gibbins
Chief Executive

Date Formed No. staff 2
Income £100,000 Expenditure £99,000

Mission Statement

Institute of Terrestrial Ecology

University College of North Wales Bangor Research Unit
Orton Building Deiniol Road
Bangor Gwynedd LL57 2UP

Tel 01248 370045
Fax 01248 355365
E Mail lsa@uk.ac.nerc-bangor.uni

Contact Lloyd Anderson

Date Formed 1974 No. staff 280
Income £11.8 million Expenditure £12 million

Mission Statement

The Institute will develop long term, multi-disciplinary research and exploit new technology to understand the science of the natural environment, with particular emphasis on terrestrial ecosystems.

Intermediate Technology

Myson House Railway Terrace Rugby CV21 3HT

Tel 01788 560631
Fax 01788 540270
E Mail itdg@gn.apc.org

Contact John Twigg

Date Formed 1965 No. staff 80 UK, 250 o/s
Income £7 million Expenditure £7 million

Mission Statement

Intermediate Technology enables poor people in the South to develop and use skills and technologies which give them more control over their lives and which contribute to the sustainable development of their communities.

International Alert

1 Glyn Street London SE11 5HT

Tel 0171 793 8383
Fax 0171 793 7975
E Mail intlalert@gn.apc.org

Contact Kumar Rupesinghe

Date Formed 1985 No. staff 40
Income £1.7 million Expenditure £1.5 million

Mission Statement

International Alert plays an active role in worldwide networks which work to prevent and resolve violent conflict within countries and to reconcile parties at war with each other through mediation, facilitation and education.

International Christian Relief

PO Box 180 16 St Johns Hill Sevenoaks
Kent TN13 3NP

Tel 01732 450250
Fax 01732 741190
E Mail

Contact Terri Lewis
Assistant to National Director

Date Formed 1978 No. staff 3
Income Expenditure

Mission Statement
To bring aid and relief to those in the third world.

International Extension College

Dale's Brewery Gwydir Street Cambridge CB1 2LJ

Tel 01223 353321
Fax 01223 464734
E Mail

Contact John Thomas

Date Formed 1984 No. staff 20
Income £1.06 million Expenditure £1.05 million

Mission Statement

IEC promotes development and improvement in the quality of life through the expansion of educational opportunities, using distance education techniques. IEC works with refugees and displaced persons.

International Institute for Environment and Development (IIED)

3 Endsleigh Street London WC1H 0DD

Tel 0171 388 2117
Fax 0171 388 2826

E Mail humansiied@gn.apc.org

Contact David Satterthwaite
Director Human Settlements

Date Formed 1971   No. staff 50
Income £3.4 million   Expenditure £3.4 million

Mission Statement

IIED seeks to promote sustainable patterns of development through research, policy studies, consensus-building and public information.

International NGO Training and Research Centre (INTRAC)

P.O. Box 563 Oxford OX2 6RT

Tel 01865 201851
Fax 01865 201852
E Mail

Contact Brian Pratt
Executive Director

Date Formed 1991   No. staff 6
Income £231 678   Expenditure £208 235

Mission Statement

Management training for NGOs in relief and development, research and consultancy.

International Rescue Corps (IRC)

8 Kings Road Grangemouth Stirlingshire Scotland FK3 9BB

Tel 01452 722251
Fax 01324 666130
E Mail

Contact John Holland

Date Formed   No. staff
Income £42 516   Expenditure £29 486

Mission Statement

The IRC is a registered voluntary search and rescue team, attending to natural and man-made disasters.

International Seismological Centre

Piper's Lane Thatcham Newbury RG13 4NS

Tel 01635 861022
Fax 01635 872351
E Mail dmcg@ib.rl.ac.uk

Contact A A Hughes
Director

Date Formed 1964   No. staff 5
Income £354 000   Expenditure £366 000

Mission Statement

The collection and analysis of world earthquake information used in studies of earthquake mechanism, earth structure and seismic hazards.

Karuna Trust

186 Cowley Road Oxford OX4 1VE

Tel 01865 728794
Fax 01865 792941
E Mail

Contact Peter Joseph
Director

Date Formed 1980   No. staff 7
Income £500 000   Expenditure

Mission Statement

Social and cultural work among former untouchables and other deprived groups in India.

Living Earth

Warwick House 106 Harrow Road London W2 1XD

Tel 0171 258 1823
Fax 0171 258 1824
E Mail livearth@gn.apc.org

Contact Roger Hammond

Date Formed 1987   No. staff 19
Income £400 000   Expenditure £400 000

Mission Statement

Work with youth and industry for protection of natural resources and promotion of their sustainable use through environmental education.

Llewelyn Davies Planning

Brook House 2-16 Torrington Place London WC1E 7HN

Tel 0171 637 0181
Fax 0171 637 8740
E Mail

Contact Jon Rowland

Date Formed 1963 *No. staff* 60
Income *Expenditure*

Mission Statement

Lloyd's Register (Civil and Structural Engineering)

29 Wellesley Road Croyden Surrey CR0 2AJ

Tel 0181 6814764
Fax 0181 6816814
E Mail

Contact J R Magurie

Date Formed 1760 *No. staff* 3500
Income £190 million *Expenditure* £190 million

Mission Statement

To ensure high standards of safety on land and at sea.

London School of Hygiene & Tropical Medicine

Keppel Street London WC1E 7HT

Tel 0171 927 2464
Fax 0171 580 7593
E Mail B.Judge@lshtm.ac.uk.

Contact Barbara Judge
Project Officer

Date Formed 1985 *No. staff* 269
Income £25.5 million *Expenditure* £24.3 million

Mission Statement

To contribute to the improvement of health worldwide through the pursuit of excellence in research, postgraduate teaching, advanced training and consultancy in international public health and tropical medicine.

Marie Stopes International

62 Grafton Way London W1P 5LE

Tel 0171 388 3034
Fax 0171 3834 544
E Mail

Contact Patricia Hindmarsh
Director of Campaigns

Date Formed 1973 *No staff* 30
Income £7 million *Expenditure* £7 million

Mission Statement

The prevention of unwanted births.

Medical Emergency Relief International (MERLIN)

1a Rede Place Chepstow Place London W2 4TU

Tel 0171 229 4560
Fax 0171 243 1442
E Mail

Contact Christopher Besse

Date Formed 1993 *No staff* 18
Income £2.4 million *Expenditure* £2.2 million

Mission Statement

Medical emergency relief through the use of professional and volunteer health workers and appropriate support teams.

Megacities Project

The Institution of Civil Engineers 1 Great George Street London SW1P 3AA

Tel 0171 222 7722
Fax 0171 799 1325
E Mail

Contact Mr Louis Solway
Project leader

Date Formed *No. staff*
Income *Expenditure*

Mission Statement

Mott MacDonald Group

St Anne House 20/26 Wellesley House Croyden Surrey CR9 2UL

Tel 0181 6865041
Fax 0181 6815706
E Mail mm-croy:cpe

Contact C P Ellinas
Director

Date Formed 1989 *No staff* 4 500
Income £ 160 million *Expenditure* £ 150 million

Mission Statement

National Rivers Authority

Rivers House Waterside Drive Aztec West Almondsbury Bristol BS12 4UD

Tel   01454 624400
Fax   01454 624409
E Mail

Contact   Kevin Bond
           Director of Operations

Date Formed 1989   No staff 7 500
Income   £400 million   Expenditure

Mission Statement

To protect and maintain the water environment.

Natural Resources Institute

Resource Management Division Central Avenue Chatham Maritime Chatham Kent ME4 4TB

Tel   01634 880 088
Fax   01634 880 066/77
E Mail   nri@ukc.ac.uk

Contact   Paul Hindmarsh
           Head Food Storage Management

Date Formed 1987   No staff   450
Income   £32 million Expenditure   £31.1 million

Mission Statement

To alleviate poverty and hardship in developing countries by working with them to increase the productivity of their renewable natural resources in a sustainable way.

NERC Institute of Hydrology

Maclean Building Crowmarsh Giddord Wallingford Oxfordshire OX10 8BB

Tel   01491 838800
Fax   01491 692424
E Mail

Contact

Date Formed   No staff
Income   Expenditure

Mission Statement

NERC Unit for Thematic Information Systems (NUTIS)

Dept of Geology University of Reading Reading RG6 2AB

Tel   01746 875123 X7764
Fax   01746 755865
E Mail   gw@mail.nerc-nutis.ac.uk

Contact   Geoff Wadge

Date Formed 1983   No staff 15
Income   Expenditure

Mission Statement

Research into the environmental sciences.

Noble Denton Weather Services Ltd

Noble House 131 Aldersgate Street London EC1A 4EB

Tel   0171 606 4961
Fax   0171 606 5035
E Mail   nlynagh@cix.compulink.co.uk

Contact   Norman Lynagh
           Managing Director

Date Formed 1986   No staff 14
Income   £750 000   Expenditure

Mission Statement

Our aim is to be the leading meteorological and oceanographic consultancy with a worldwide capability. We are not targeting disasters as a core business but we are often called upon to provide accurate meteorological information in the event of a major environmental disaster such as a land-falling, hurricane or severe flooding.

Opportunity Trust

PO Box 642 Oxford OX1 4DZ

Tel   01865 794 411
Fax   01865 295 161
E Mail   impact@opportuk.demon.co.uk

Contact   William day

Date Formed 1991   No staff 8
Income   Expenditure

Mission Statement

Creating income opportunities with the poor.

Overseas Development Administration (Emergency Aid Department)

Overseas Development Administration 94 Victoria Street London SW1E 5JL

Tel	0171 917 7000
Fax	0171 917 0502
E Mail	druojed.us3@oda.gnet.gov.uk
Contact	Janet Douglas Head of Section

Date Formed		No staff	2
Income	£2 million	Expenditure	£1.9 million

Mission Statement

To provide material aid, including food, medical care, shelter and personnel, and finance, and advise to save and preserve lives during emergency situations and in the immediate post-emergency rehabilitation phase; also to cope with population displacements arising out of emergencies. ODA objectives are: to support economic reform, to enhance productive capacity, to help achieve good government, to finance activities directly benefiting poor people, to promote human development, including better education and health, and children by choice; to promote the status of women; and to help tackle environmental problems.

Ove Arup Partnership

13 Fitzroy Street London W1P 6BQ

Tel	0171 636 1531
Fax	0171 465 2150
E Mail	
Contact	Ed Booth Richard Hughes

Date Formed	1946	No staff	3 500 w/w
Income		Expenditure	

Mission Statement

Achieving excellence in engineering design.

Overseas Development Group

School of Development Studies University of East Anglia Norwich NR4 7TJ

Tel	01603 457880
Fax	01603 505262
E Mail	
Contact	David Leddon Managing Director

Date Formed	1980	No	
Income		Expenditure	

Mission Statement

Overseas Deve...

Regents College Inner Circle Rege. London NW1 4NS

Tel	0171 487 7413
Fax	0171 487 7590
E Mail	odi@gn.apc.org
Contact	John Borton Research Fellow

Date Formed		No staff	
Income		Expenditure	

Mission Statement

Founded in 1960 as an independent centre for development research and a forum for discussion of the problems facing developing countries.

Oxfam (Emergency Department)

274 Banbury Road Oxford OX2 7DZ

Tel	01865 311311
Fax	01865 312224
E Mail	
Contact	Marcus Thompson Emergencies Director

Date Formed	1942	No staff	
Income		Expenditure	

Mission Statement

To relieve poverty, distress and suffering throughout the world, to educate people about the causes and effects of poverty, distress and suffering, and to campaign for a world without them.

Oxford Centre for Disaster Studies (OCDS)

PO Box 137 Oxford OX4 1BB

Tel	01865 202772
Fax	01865 202 848
E Mail	100612.1153@compuserve.com
Contact	Ian Davis

Date Formed	1993	No staff	4
Income		Expenditure	

Mission Statement

The OCDS aims to reduce risk and improve the quality of response following disasters within the context of longer term sustainable development.

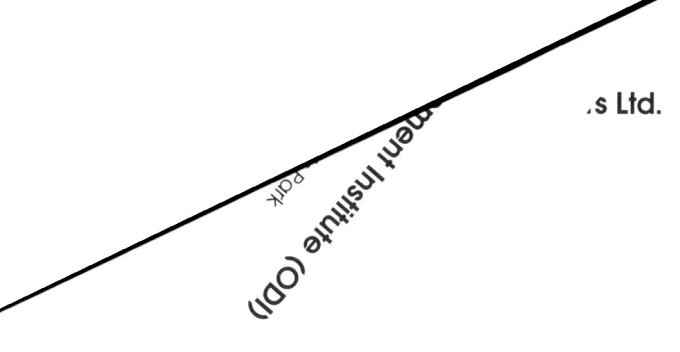

Da...		No staff	81
Incom... ...llion		Expenditure	£78 million

Mission Statement

To carry out site consulting engineering investigations and analyses; to design works (domestic, commercial and industrial buildings, highways, bridges, water-supply and waste-water schemes); construction management.

PLAN International

Chobham House Christchurch Way Woking Surrey GU21 1JG

Tel   01483 755 55
Fax   01483 756 505
E Mail   planih@gn.apc.org

Contact   Hans van der Oosten
            Programme Manager

Date Formed   1937   No staff
Income   Expenditure

Mission Statement

Plan International strives to achieve lasting improvements in the quality of life of deprived children in developing countries.

Post-War Reconstruction & Development Unit (PRDU)

University of York The King's Manor York YO1 2EP

Tel   01904 433 959
Fax   01904 433 949

E Mail   IAAS1@York.ac.uk

Contact   Sultan Barakat
            Director

Date Formed   1992   No staff 2
Income   Expenditure

Mission Statement

PRDU aims to provide an accessible and professionally relevant multi-disciplinary training on issues of disaster intervention and post-war reconstruction planning and management.

Quaker Peace and Service

Friends' House Euston Road London NW1 2BJ

Tel   0171 387 3601
Fax   0171 388 1977
E Mail   qps1@gn.apc.org

Contact   Andrew Clark
            General Secretary

Date Formed   1815   No staff   20 UK, 20 o/s
Income   £1.5 million   Expenditure   £1.5 million

Mission Statement

QPS work related to disasters includes mitigation and preparedness through the Quaker witness to the prevention of violent conflict and the reconciliation of those who are or have been violently opposed to each other.

Refugee Studies Programme

Queen Elizabeth House University of Oxford 21 St Giles Oxford OX1 3LA

Tel   01865 270 722
Fax   01865 270 721
E Mail   RSP@qeh.ox.ac.uk

Contact   Barbara Harrell-Bond
            Director

Date Formed   1982   No staff   18
Income   £947 075   Expenditure   £771 000

Mission Statement

The Refugee Studies Programme aims to increase understanding of the causes, consequences and experiences of forced migration, through multi-disciplinary research, teaching, publications, seminars and conferences.

Registered Engineers for Disaster Relief (REDR)

1-7 Great George Street London SW1P 3AA

Tel   0171 233 3116
Fax   0171 222 0564
E Mail

Contact   David P Ede

Date Formed 1979   No   5
Income   £165 906   Expenditure   £168 140

Mission Statement

RedR is a registered charity providing carefully selected engineers and technicians to relief organisations world wide to enhance their ability to save lives in disasters.

Responding To Conflict

Woodbrooke College 1046 Bristol Road
Birmingham B29 6LJ

Tel 0121 415 5641
Fax 0121 415 4119
E Mail

Contact Simon Fisher
Director

Date Formed 1890   No staff 3
Income   Expenditure

Mission Statement

The sharing of conflict-handling skills and experience to assist organisations working for development, human rights and peace in situations of instability and conflict, through practical training and consultancy.

Reynolds Geo-Sciences Ltd

10 Bron y Nant Mold Clwyd
North Wales CH7 1UX

Tel 01352 755 998
Fax 01352 755 998
E Mail

Contact Dr John Reynolds

Date Formed 1994   No staff 1
Income   Expenditure

Mission Statement

To provide high quality technical consultancy services cost effectively to host organisations in order to establish and develop local expertise and resources; to assess the risk of potential goehazards, vulnerability of local facilities, and to initiate and monitor mitigation measures.

Rural Resources Management Ltd

PO Box 144 Bourn Cambridge CB3 7SL

Tel 01954 718 026
Fax 01954 718 027
E Mail

Contact Steve Jones

Date Formed 1982   No staff 2
Income   Expenditure

Mission Statement

A consultancy specialising in participatory rural development, including disaster preparedness, mitigation and management.

Saferworld

3rd Floor 34 Alfred Place London WC1E 7DP

Tel 0171 580 8886
Fax 0171 631 1444
E Mail

Contact Hugh Venables
Executive Director

Date Formed 1989   No staff 12
Income   Expenditure

Mission Statement

To investigate common security approaches to global problems. Saferworld specifically addresses the arms trade, conflict prevention and management and the true cost of conflict.

Save The Children Fund UK

17 Grove Lane Camberwell London SE5 8RD

Tel 0171 703 5400
Fax 0171 793 7626
E Mail

Contact John Seaman
Director Overseas Development

Date Formed 1919   No staff 4000 o/s
Income £113 million   Expenditure

Mission Statement

SCF aims are to achieve 'The Rights of the Child', now the International Convention on the Rights of the Child.

School of Development Studies, UEA

University of East Anglia Norwich NR4 7TJ

Tel 01603 593 376
Fax 01603 765 741
E Mail

Contact John Cameron

Date Formed 1982   No staff 33
Income   Expenditure

Mission Statement

Development teaching, research and consultancy.

School of Humanities, University of Greenwich

University of Greenwich Wellington Street Greenwich London SE18 6PF

Tel　　　　0181 331 8805
Fax
E Mail

Contact　　Terry Cannon

Date Formed 1977　　*No staff*
Income　　　　　　　*Expenditure*

Mission Statement

University research, training and teaching.

School of Oriental and African Studies (SOAS)

Thornhaugh Street Russell Square London WC1H 0XG

Tel　　　0171 323 6159
Fax　　　0171 436 3844
E Mail　　ta1@soas.ac.uk

Contact　　J A Allan

Date Formed 1994　　*No staff* 1 full-time
Income　　£12 000　　*Expenditure* £12 000

Mission Statement

The group at SOAS is concerned with water and food security, with the development of geographical information systems incorporating a range of data including earth observation (remote sensoring).

Scottish Catholic International Aid Fund (SCIAF)

5 Oswald Street Glasgow G1 4QR

Tel　　　0141 221 4447
Fax　　　0141 221 2373
E Mail　　(Geonet) Geo2: SCIAF Scotland

Contact　　Joan Mcerlean
　　　　　　Emergencies Officer

Date Formed 1965　　*No staff* 15
Income　　£2.5 million　　*Expenditure*

Mission Statement

SCIAF is the overseas agency of the Catholic Church in Scotland. Its vision of development is to empower the poor and oppressed to become agents of change for the benefit of themselves, their communities and the world as a whole. SCIAF supports relief and disaster development projects in Africa, Asia and Latin America. SCIAF conducts development education in Scotland and campaigns for a more just world.

Sedgwick Global
(Part of Sedgwick Group)

Bristol and West House 173 Friar Street Reading RG1 1BP

Tel　　　01734 585 235
Fax　　　01734 585 572
E Mail

Contact　　C T K Toomer
　　　　　　Assistant Director/Risk Consultant

Date Formed 1884　　*No staff* 15 400 o/s
Income £900 million　*Expenditure* £800 million

Mission Statement

Our aim is to meet our clients' needs by giving them quality service. To make this happen, we know that Sedgwick's business is not only about broking, it's about advice. We know that to gain and retain our clients' trust we must address their needs individually; have a true understanding of their businesses; use the breadth and depth of our skills and resources to provide them with well-considered and innovative solutions; and deliver those solutions expertly and cost effectively.

Silsoe Research Institute

Wrest Park Silsoe Bedford MK45 4HS

Tel　　　01525 860000
Fax　　　01525 860156
E Mail　　derek.sutton@bbsrc.ac.uk

Contact　　Derek Sutton
　　　　　　Head, Overseas Division

Date Formed 1960　　*No staff* 280
Income £9 million　　*Expenditure*

Mission Statement

The work of the overseas division is most relevant in this instance and can be described as research and development in engineering and the physical sciences for agriculture and rural development.

SOS Sahel

1 Tolpuddle Street London N1 0XT

Tel　　　0171 837 9129
Fax　　　0171 837 0856
E Mail　　sossaheluk@gn.apc.org

Contact　　Duncan Fulton
　　　　　　Senior Programme Coordinator

Date Formed 1981　　*No staff* 8
Income £1.6 million　*Expenditure* £1.6 million

Mission Statement

Natural resources management, dryland forest, water resources and dryland agriculture.

Support Action for Emergencies (SAFE)

Cumbers Liss Hampshire GU33 7LL

Tel 01730 892167
Fax 01730 895232
E Mail

Contact Simon Kendall, Director

Date Formed 1995 No staff
Income Expenditure

Mission Statement

To enable the Christian Church world-wide to respond to disasters and emergencies through appropriate preplanning including education, training and other measures; to facilitate effective recovery actions to reduce and mitigate their impact; and to promote viable development.

Systems Group Faculty of Technology, The Open University

Walton Hall Milton Keynes Beds MK7 6AA

Tel 01908 653628
Fax 01908 652175
E Mail

Contact V Bignell
Failures Chair

Date Formed 1974 No staff
Income Expenditure

Mission Statement

To add to the understanding of systems failures.

Teaching Aids at Low Cost (TALC)

P.O. Box 49 St Albans Herts AL1 4AX

Tel 01727 853 869
Fax 01727 846 852
E Mail

Contact Indira Benbow
Administrator

Date Formed No staff
Income Expenditure £250 000

Mission Statement

Providing teaching aids

Tear Fund

100 Church Rd Teddington Middx TW11 8QE

Tel 0181 977 9144
Fax 0181 943 3594
E Mail

Contact Michael Wall
Disaster Response Manager

Date Formed 1968 No staff 220
Income £18 million Expenditure £18 million

Mission Statement

To enable indigenous organisations to provide relief and development assistance in the name of Christ.

The Centre for Crisis Psychology

Four Arches Broughton Hall Skipton
Skipton North Yorks BD23 3AE

Tel 01756 796383
Fax 01756 796384
E Mail

Contact Barbara Wright
Administration and Marketing

Date Formed 1988 No staff 6
Income Expenditure

Mission Statement

Trauma treatment and aftercare.

The Research Centre

University of Luton 24 Crawley Green Road
Luton Beds LU1 3LF

Tel 01582 456 843
Fax 01582 459 787
E Mail sellis@uk.ac.luton.vax2

Contact Sue Ellis

Date Formed 1992 No staff 1
Income Expenditure

Mission Statement

Investigation and analysis of shelter requirements for refugees and displaced persons in the aftermath of war.

The Royal Society

6 Carlton House Terrace London SW1Y 5AG

Tel 0171 839 5561
Fax 0171 930 2170
E Mail ezrs001@ulcc.ac.uk

Contact Ruth Cooper
Science Advice Section

Date Formed 1660 *No staff* 120
Income *Expenditure* £25 million

Mission Statement

The Royal Society is an independent academy promoting the natural and applied sciences, nationally and internationally and is the national academy of sciences for the UK.

The Society for Earthquake and Civil Engineering Dynamics (SECED)

c/o The Institute of Civil Engineers
1-7 Great George Street London SW1P 3AA

Tel 0171 222 7722
Fax 0171 799 1325
E Mail

Contact Mary Kinsella
Secretary
Date Formed *No staff*
Income *Expenditure*

Mission Statement

SECED's interests embrace earthquake engineering as well as the wider sub-disciplines of civil engineering dynamics, including blast, impact and other vibration problems.

Trocaire

169 Booterestoon Avenue Blackrock Dublin Ireland

Tel 288 5385
Fax 283 6022
E Mail Niall@Trocaire.ie.

Contact Niall Toibin
Emergency Co-ordinator

Date Formed 1973 *No staff* 60
Income £14 million *Expenditure* £10 million

Mission Statement

To assist those in need in developing countries and to make Irish people more aware of these needs and our duty in justice towards them.

UK Committee for UNICEF

55 Lincoln Inn Fields London WC2 3NB

Tel 0171 405 5592
Fax 0171 405 2332
E Mail

Contact Robert Smith
Executive Director

Date Formed 1956 *No staff* 50
Income £14 million *Expenditure* £2.5 million

Mission Statement

To further the charitable work throughout the world of the United Nations Children's Fund (UNICEF) for the relief of children who are poor or in need of care and attention, the education and training of children, the promotion of the health of children. The UK Committee for UNICEF is a UK registered charity and the emergency staff are managed by the office of Emergency Programmes, New York. The UK Committee has an agreement to represent UNICEF in the UK.

UK National Co-ordination Committee for the IDNDR

The Royal Academy of Engineering 29 Great Peter Street London SW1P 3LW

Tel 0171 222 2688
Fax 0171 233 0054
E Mail

Contact Miss Jacqueline Baines
Secretary to the Committee

Date Formed 1991 *No staff*
Income *Expenditure*

Mission Statement

Co-ordinating responsibility for the UK IDNDR activities and encouragement of IDNDR related activities.

UK- Med

North Staffordshire Royal Infirmary Princes Road Hartshill Stoke-on-Trent ST4 7LN

Tel 0178 274 9722
Fax 0161 973 6276
E Mail

Contact A D Redmond

Date Formed *No staff* 3

Income £1.82 million *Expenditure* £1.5 million

Mission Statement

To support international emergency medical aid.

United Kingdon Jewish Aid

33 Seymour Place London W1H 6AT

Tel	0171 723 3442
Fax	0171 723 3445
E Mail	
Contact	Michael Harris
Director

Date Formed	1989	No staff	1
Income	£215 000	Expenditure	£160 000

Mission Statement

UK Jewish Aid is a registered charity that works to counteract poverty, sickness and deprivation without regard to race, religion or colour. Founded in 1989 it channels the UK Jewish Community's contribution to non-sectarian relief and development world wide. UKAID aims to: respond to international emergencies, run and assist in long-term development programmes, enable people to become self-reliant and self sufficient.

United Nations Association (UNA)

3 Whitehall Court London SW1A 2EL

Tel	0171 930 2931
Fax	0171 930 5893
E Mail	
Contact Name	Malcolm Harper
Director

Date Formed	1986	No staff	15
Income	£300 000	Expenditure	£300 000

Mission Statement

Promotion of wider understanding of and support for the United Nations.

Volcano Geophysics Group

Dept. of Earth Sciences Open University Walton Hall Milton Keynes MK7 6AA

Tel	01908 653 949
Fax	01908 655 151
E Mail	
Contact	Hazel Rymer

Date Formed		No staff	5 full-time
Income	£300 000	Expenditure	

Mission Statement

Water Engineering and Development Centre (WEDC)

Loughborough University of Technology
Loughborough Leics LE11 3TU

Tel	01509 222 628
Fax	01509 211 079
E Mail	
Contact	R Reed
Programme Manager

Date Formed	1985	No staff	20
Income	£1 million	Expenditure	£950 000

Mission Statement

Education, training, research and consultancy for the planning provision and management of physical infrastructure for development in low and middle income countries.

Wind Engineering Society

c/o Institution of Civil Engineers 1 Great Peter Street London SW1P 3AA

Tel	0171 222 7722
Fax	0171 799 1325
E Mail	
Contact	Mary Kinsella
Secretary to WES Executive

Date Formed	1990	No staff	
Income	£5000	Expenditure	£5000

Mission Statement

To provide a service to the community related to the understanding and solution of wind engineering problems.

World Association for Disaster and Emergency Medicine

Accident and Emergency Dept Royal London Hospital Whitechapel London E1 1BB

Tel	0171 377 7000
Fax	0171 377 7014
E Mail	
Contact	Judith Fisher
Honorary Secretary

Date Formed	1975	No staff	1 part time
Income		Expenditure	

Mission Statement

To help make effective disaster and emergency medicine a worldwide reality. The Association meets as a body every 2 years and has about 500 members.

World Conservation Monitoring Centre

219 Huntingdon Road
Cambridge CB3 0DL

Tel 01223 277 314
Fax 01223 277 136
E Mail

Contact   Richard Luxmoore

Date Formed 1988   *No staff* 50

Income £1.8 million   *Expenditure* £1.8 million

Mission Statement

WCMC provides information services on the conservation and sustainable use of species and ecosystems and supports others in the development of their own information management services.

World Development Movement

25 Beehive Place London SW9 7QR

Tel 0171 737 6215
Fax 0171 274 8232
E Mail

Contact   Roger Briottet
           Director

Date Formed   *No staff* 14
Income £504 723   *Expenditure* £481 942

Mission Statement

World Development Movements campaigns are mainly aimed at seeking justice for the world's poor.

World Vision UK

599 Avebury Boulevard Central Milton Keynes
Bucks MK9 3PG

Tel 01908 841 000
Fax 01908 841 015
E Mail

Contact   Susan Barber
           Manager Africa Region/Department

Date Formed 1982   *No staff* 60
Income £16.6 million   *Expenditure* £13 million

Mission Statement

To work with the poor for transformation and justice.

Y Care International

640 Forest Road
London E17 3DZ

Tel 0181 520 5599
Fax 0181 503 7461
E Mail

Contact   Mr Stephen Stordy
           Overseas Projects Manager

Date Formed 1984   *No staff* 12
Income £ 3. 3 million   *Expenditure* £1.5 million

Mission Statement

Y Care International works in partnership with young people world-wide, in the light of the Christian commitment of the YMCA to help them enrich their own lives, to build a more just world free from poverty and to provide appropriate disaster and emergency relief.

UK IDNDR focal points

The following persons have kindly agreed to act as focal points: to answer general and technical queries or where matters are not in their area of expertise to deflect enquiries to more suitable persons.

The following list of focal points is being distributed widely, with the audit, to various government Departments, EU, UN agencies, contracting bodies as well as media. The names have been carefully selected and approved by the UK National IDNDR Co-ordination Committee. All persons on this list are prepared to provide comment and broad advice at no cost. Where possible two names have been provided, since individual focal points may not always be available.

It should be emphasised that any advice offered by these focal points comes from them personally, based on their experience and expertise; it does not necessarily reflect the views of the organisation they work for or the UK IDNDR National Co-ordination Committee.

This section comprises the following headings:

- General knowledge of disaster preparedness/mitigation;
- Hazard types and related sectors;
- Sectors;
- Country knowledge;
- Government focal points;
- NGO focal points;
- International focal points.

General knowledge of disaster preparedness/mitigation

John Borton	ODI, Regents College, Inner Circle, Regents Park, London, Tel: 0171 917 1000, Fax: 0171 917 1010
Ian Davis	OCDS, P. O. Box 137, Oxford, OX4 1BB, Tel: 01865 202 772, Fax: 01865 202 848
Tony Eades	(Queries regarding UK IDNDR activities) Secretary UK National Co-ordination Committee The Royal Academy of Engineering, 29 Great Peter Street, Westminster, London, SW1P 3DL, Tel: 0171 222 2688, Fax: 0171 233 0054
Mike Hulme	(General enquires regarding drought and climate change) Climatic Research Unit, University of East Anglia, Norwich, NR4 7TJ, Tel: 01603 592 088, Fax: 01603 507 784
Ken Westgate	Disaster Preparedness Centre (DPC), Cranfield University, RMCS, Shrivenham, Swindon, SN6 8LA, Tel: 01793 785 287, Fax: 01793 782 179

Hazard types and related sectors

Complex emergencies

General

Hugo Slim | CENDEP (Centre for Development and Emergency Planning), Oxford Brookes University, Gypsy Lane, OX3 0BP, Oxford, Tel: 01865 483413, Fax: 01865 483298

Peacekeeping, the role of the military

Mark Duffield | School of Public Policy, University of Birmingham, Edgbaston, Birmingham, B15 2TT, Tel: 0121 414 5021, Fax: 0121 414 3279

Planning & management of refugee settlements

Dr Roger Zetter | School of Planning, Oxford Brookes University, Gipsy Lane, Oxford, OX3 0BP, Tel: 01865 483 450, Fax: 01865 483 559

Disease and epidemics

Dr A. Redmond | UK-Med, Windsor House, North Staffordshire Hospital, Princes Road, Hartshill, Stoke-On-Trent, ST4 7LN, Tel: 01782 749722, Fax: 01782 749722

Epidemiology of disease

Dr Peter Baxter | Department of Community Medicine, Fenners Gresham Road, Cambridge, CB1 2ES, Tel: 01223 336 590, Fax: 01223 336 584

Drought/famine

Physical aspects

Dr J. S. Wallace | Institute of Hydrology, Crowmarsh Gifford, Wallingford, OX10 8BB, Tel: 01491 838 800, Fax: 01491 838 097

Integration of natural and social factors

Dr T. Downing | Environmental Change Unit, 1a Mansfield Road, Oxford, OX1 3TB, Tel: 01865 281 180, Fax: 01865 281 181

Early warning systems
(macro level socio-economic analysis)

Dr J. Holt | Save the Children Fund, Mary Datchelor House, 17 Grove Lane, Camberwell, London, SE5 8RD, Tel: 0171 703 5400, Fax: 0171 793 7626

Measures to reduce drought impact
(crop diversification/soil conservation, irrigation etc.)

Dr Wright | Centre for Arid Zone Studies, Thoday Building, University of Wales, Bangor, Gwynedd, LL57 2UW, Tel: 01248 382 346, Fax: 01248 364 717

Earthquake

General

Dr R. D. Adams	International Seismology Centre, Pipers Lane, Thatcham, Berkshire, RG19 4NS, Tel: 01635 861 022, Fax: 01635 872 351
A. B. Walker	British Geological Survey, Global Seismology Group, Murchison House, West Mains Road, Edinburgh, EH9 3LA, Tel: 0131 667 1000, Fax: 0131 667 1877

Engineering seismology

E. Booth	Ove Arup and Partners, 13 Fitzroy Street, London, W1P 6BQ, Tel: 0171 636 1531, Fax: 0171 465 2150

Reducing the impact of earthquakes on intrinsic structures

R. Hughes	Ove Arup and Partners, 13 Fitzroy Street, London, W1P 6BQ, Tel: 0171 636 1531, Fax: 0171 465 2150

Flood

General

Prof B. Wilkinson	Institute of Hydrology, Crowmarsh Gifford, Wallingford, OX10 8BB, Tel: 01491 838 800, Fax: 01491 838 097
Prof D. Parker	Flood Hazard Research Centre, Middlesex University, Queensway, Enfield, EN34 SF, Tel: 0181 362 5359, Fax: 0181 362 5403

Flood warning systems and measures to reduce flood impact

Prof D. Parker	Flood Hazard Research Centre, Middlesex University, Queensway, Enfield, EN3 4SF, Tel: 0181 362 5359, Fax: 0181 362 5403

Hurricane/cyclone

General

Dr W. H. Lyne	Met Office, London Road, Bracknell, Berkshire, RG12 2SZ, Tel: 01344 854 901, Fax: 01344 856 909

Cyclone warning systems

Dr W. H. Lyne	Met Office, London Road, Bracknell, Berkshire, RG12 2SZ, Tel: 01344 854 901, Fax: 01344 856 909

Wind engineering

Prof B. E. Lee	Department of Civil Engineering, University of Portsmouth, Portsmouth, PO1 3QL, Tel: 01705 842 423, Fax: 01705 842 521
Dr N. J. Cook	Building Research Establishment, Garston, Watford, WD2 7JR, Tel: 01923 664 218, Fax: 01923 664 096

Landslide

General

Prof D. Brunsden	Department of Geography, Kings College, London, WC2R 2LS, Tel: 0171 873 2577, Fax: 0171 873 2287
Prof D.K.C. Jones	Department of Geography, London School of Economics, Houghton Street, London, WC2A 2AE, Tel: 0171 955 7577, Fax: 0171 955 7721

Landslide monitoring and engineering

Sir John Knill	Highwood Farm, Shaw-Cum-Donnington, Newbury, Berkshire, RG14 2TB, Tel: 01635 552 300, Fax: 01635 368 26

Volcanoes

General

Prof J. Guest	University College London, Gower Street, London, WC1E 6BT, Tel: 0181 959 0421, Fax: 0181 906 4161
Prof R.S.J. Sparks	Geology Department, University of Bristol, Bristol, BS8 1RJ, Tel: 0117 928 7789, Fax: 0117 925 3385

Volcanic/prediction warning systems

Hazel Rymer	Volcano Geophysics Group, Open University, Walton Hall, Milton Keynes, MK7 6AA, Tel: 01908 653 949, Fax: 01908 655 151

Social impact of volcanoes

Dr D. Chester	Department of Geography, University of Liverpool, Liverpool, L69 3BX, Tel: 0151 794 2876, Fax: 0151 794 2866

Volcanic preparedness and health aspects of volcanoes

Dr P. Baxter	Department of Community Medicine, Fenners, Gresham Road, Cambridge, CB1 2ES, Tel: 01223 336 590, Fax: 01223 336 584

Sectors

Agriculture

Hugh Brammer	37 Kingsway Court, Hove, East Sussex. BN3 2LP. Tel/Fax 01273 733 114

Anthropology

Dr David Turton	Department of Social Anthropology, University of Manchester, Brunswick Street, Manchester, M13 9PLUK, Tel: 0161 275 3999, Fax: 0161 275 3970

Appropriate technology/indigenous knowledge

John Twigg — Intermediate Technology, Myson House, Railway Terrace, Rugby, CV21 3HT, Tel: 01788 560 631, Fax: 01788 540 270

Building and architecture

Dr Ian Davis — OCDS, P.O. Box 137, Oxford, OX4 1BB, Tel: 01865 202 772, Fax: 01865 202 848

Communications/information management

Dr R. Stephenson — 22 Sunderland Road, Ealing, London, W5 4JY, Tel: 0181.5676034, Fax: 0181.5665971

Conflict prevention/tension reduction

Hugo Slim — CENDEP(Centre for Development and Emergency Planning), Oxford Brookes University, Gipsy Lane, Oxford, OX3 0BP. Tel: 01865 483 413, Fax: 01865 483 298

Development economics

Prof Piers Blaikie — Overseas Development Group, University of East Anglia, Norwich, NR4 7TJ, Tel: 01603 457 880, Fax: 01603 505 262

Emergency management

Eric Alley — Farthings, Gembling, Driffield, East Yorkshire, YO25 8HS, Tel: 01262 488 518, Fax: 01262488 518

Peter Burton — Head of Disaster Unit, Emergency Aid Department, ODA, 94 Victoria Street, London, SW1E 5JL, Tel: 0171 917 0470, Fax: 0171 917 0502

Engineering

Prof Brian Lee — High Wind, Dept. of Civil Engineering, University of Portsmouth, Portsmouth, PO1 3QL, Tel: 01705 842423, Fax: 01705 842521

Prof Amr El-Nashai — Civil Engineering Department, Imperial College, South Kensington, London, SW7 2BU, Tel: 0171 589 5111, Fax: 0171 594 6053

Food Security

M. Buchanan-Smith — Head of Emergency Department, Action Aid, Hamlyn House, MacDonald Road, Archway, London, N19 5PG, Tel: 0171 2814101, Fax: 0171 281 2076

Geomorphology

Prof Brunsden — Department of Geography, Kings College, London, WC2R 2LS, Tel: 0171 873 2577, Fax: 0171 873 2287

Health/epidemiology/nutrition/medicine

Dr A. Redmond — UK-Med, Windsor House, North Staffordshire Hospital, Princes Road, Harthill, Stoke-on-Trent, ST4 7LN, Tel: 01782 749722, Fax: 01782 749 722

Dr M. Kapila — Humanitarian Aid Adviser, Emergency Aid Apartment, ODA, 94 Victoria Street, London, SW1 5JL, Tel: 0171 917 1000

Hydrology

Prof B. Wilkinson — Institute of Hydrology, Crowmarsh Gifford, Wallingford, OX10 8BB, Tel: 01491 838 800, Fax: 01491 838 097

Insurance/Reinsurance

David Whiting — Alexander Howden Group Ltd, Tel: 0171 216 3062, Fax: 0171 626 1648

Commercial Risk Management

Charles Toomer — Sedgwick Global (Worldwide), Bristol West House, 173 Friar Street, Reading, RG1 1BP, Tel; 01734 654224, Fax: 01734 585572

Meteorology

Dr W. H. Lyne — Met Office, London Road, Bracknell, Berkshire, RG12 2SZ, Tel: 01344 854 901, Fax: 01344 856 909

Media

Nick Cater — c/o The Red Cross, 9 Grosvenor Crescent, London, SW1X 7EJ, Tel: 01458 251 727, Fax: 01458 251 749

Military

G. Ritchie/ Mike Evans — Disaster Preparedness Centre, Cranfield University RMCS, Shrivenham, Swindon, SN6 8LA, Tel: 01793 785287, Fax: 01793 782179

Physical planning

Dr Roger Zetter — School of Planning, Oxford Brookes University, Gipsy Lane, Oxford, OX3 0BP, Tel: 01865 483 450, Fax: 01865 483 559

Remote sensing

G. Wadge — NUTIS, Department of Geography, University of Reading, RG6 6AB, Reading, Tel: 01734 875123, Fax: 01734 755865

Seismology

Dr R. D. Adams — International Seismology Centre, Pipers Lane, Thatcham, Berkshire, RG19 4NS, Tel: 01635 861022, Fax: 01635 872351

Social science research

Dr D. Turton — Department of Social Anthropology, University of Manchester, Brunswick Street, Manchester, M13 9PLUK, Tel: 0161 275 3999, Fax: 0161 275 3970

Telecommunications

Richard W. Morgan — Sedgwick Global Telecommunications, 173 Friar Street, Reading, Berkshire, RG1 1BP, Tel: 01734 585 235, Fax: 01734 585 572

Transport

Mike Evans — Disaster Preparedness Centre, Cranfield University, RMCS, Shrivenham, Swindon, SN6 8LA, Tel: 01793 785 287, Fax: 01793 782 179

Training

Ken Westgate — Disaster Preparedness Centre, Cranfield University, RMCS, Shrivenham, Swindon, SN6 8LA, Tel: 01793 785 287, Fax: 01793 782 179

Volcanology

Prof J. Guest — University College London, Gower Street, London, WC1E 6BT, Tel: 0181 959 0421, Fax: 0181 906 4161

Country knowledge

Bangladesh

Hugh Brammer — 37 Kingsway Court, Hove, East Sussex, BN3 2LP, Tel: 01273 733 114, Fax: 01273 733 114

Ethiopia

Dr Julius Holt — Save the Children Fund, Mary Datchelor House, 17 Grove Lane, Camberwell, London, SE5 8RD Tel: 0171 703 5400, Fax: 0171 793 7626

Hong Kong

Sir John Knill — Highwood Farm, Shaw-cum-Donnington, Newbury, Berkshire, RG14 2TB, Tel: 01635 552 300, Fax: 01635 368 26

India

George Ritchie — Disaster Preparedness Centre, Cranfield University, RMCS, Shrivenham, Swindon, SN6 8LA Tel: 01793 785 287, Fax: 01793 782179

Iran

Sir John Knill — Highwood Farm, Shaw-cum-Donnington, Newbury, Berkshire, RG14 2TB, Tel: 01635 552300, Fax: 01635 36826

Jordan

Dr Sultan Barakat — Post-War Reconstruction and Development Unit, IAAS, The King's Manor, University of York, York, YO1 2EP, Tel: 01904 433959, Fax: 01904 433949

Central and South America

Kevin McKemey — 14 Andrews Road, Earley, Reading, Berks, RG6 2PJ, Tel: 01734 264281, Fax: 01734 264 281

Sudan

M. Mukhier — Disaster Preparedness Centre, Cranfield University, RMCS, Shrivenham, Swindon, SN6 8LA Tel: 01793 785287, Fax: 01793 782179

Tanzania

Ken Westgate — Disaster Preparedness Centre, Cranfield University, RMCS, Shrivenham, Swindon, SN6 8LA, Tel: 01793 785287, Fax: 01793 782179

Uganda

Ken Westgate — Disaster Preparedness Centre, Cranfield University, RMCS, Shrivenham, Swindon, SN6 8LA, Tel: 01793 785287, Fax: 01793 782179

Government Focal Points

Overseas development administration (ODA)

Desks and Development Divisions:

Southern Africa

Mr Gareth Davis ODA Harare Office

Dhaka

Mr Eamoinn Taylor — First Secretary, Aid Management Office, Dhaka

East Africa

Mr Michael Ellis — Horn of Africa, East Africa Department

Southern/Central Africa

Mr Geoff Leader — Angola, Mozambique-Central and Southern Africa Department

Advisors

Guy Templer	Social development/emergency aid
Dr Mukesh Kapila	Emergency aid advisor
Mr Colin Ellis	Engineering
Dr Pene Key	Health and population
Mr John Roberts	Aid economics

NGO Focal Points

Action Aid

Margie Buchanan-Smith — Action Aid, Hamlyn House, MacDonald Road, Archway, London, N19 5PG, Tel: 0171 282 4202, Fax: 0171 281 2076

British Red Cross

Mike Adamson — British Red Cross, 9 Grosvenor Crescent, London, SW1X 7EJ, Tel: 0171 235 5454, Fax: 0171 235 0397

Intermediate Technology

John Twigg — Myson House, Railway Terrace, Rugby, CV21 3HT, Tel: 01788 560631, Fax: 01788 540270

Oxfam

Marcus Thompson — Oxfam, 274 Banbury Road, Oxford, OX2 7DZ, Tel: 01865 311311, Fax: 01865 312224

The Save the Children Fund

John Seaman — SCF, Mary Datchelor House, 17 Grove Lane, Camberwell, London, SE5 8RD, Tel: 1071 703 5400, Fax: 0171 793 7626

Tear Fund

Mike Wall — Tear Fund, 100 Church Road, Teddington, Middlesex, TW11 8QE, Tel: 0181 977 9144, Fax: 0181 977 6552

Key International Focal Points

International Decade for Natural Disaster Preparedness (IDNDR)

Olavi Elo — Director, IDNDR Secretariat, Palais des Nations, 1211 Geneva 10, Tel: 41 22 740 0377, Fax: 41 22 733 8695

Department of Humanitarian Affairs (DHA)

c/o Palais des Nations, 1211 Geneva 10, Tel: 41 22 907 1234

John Tomblin	Head of mitigation
Nahla Haidar	Disaster management training
Ola Almgren	UN disaster assessment and co-ordination
Philippa Boulle	Information/early warning

Pan American Health Organisation (PAHO)

Dr Claude de Ville de Goyet — Head of Emergencies Unit, 525 Twenty Third Street, NW, Washington DC 20037, USA, Tel: 202 861 4325, Fax: 202 775 4578

United Nations Centre for Human Settlements (UNCHS)

Ignacio Armillas Co-ordinator, Asia and Pacific Unit
P. O. Box 30030, Nairobi, Kenya
Tel: 520266, Fax: 226 473

United Nations Educational, Scientific and Cultural Organisation (UNESCO)

Bruno Haghebaert c/o CICAT (Delft University of Technology), P.O. Box 5048, 2600 GA,
Delft, The Netherlands
Fax: 00 31 15 7811791

World Health Organisation (WHO)

Dr Bassani Director of Emergency and Humanitarian Action Unit
c/o 20 Avenue Appia, 1211 Geneva 27
Tel: 41 22 791 2111, Fax: 41 22 791 0746

Directory of individuals

The following names and addresses include all those mentioned in the audit. They have been arranged as address labels to provide best possible usage.

Adams, Martin
Independent Consultant
2 Giffords Close Girton
Cambridge
CB3 0PF

Adams, Maurice
ACET
P.O.Box 3693
London
SW15 2BQ

Adams, P D
International Seismological Centre
Piper's Lane Thatcham
Newbury
RG13 4NS

Adamson, Mike
British Red Cross Society Headquarters
9 Grosvenor Crescent
London
SW1X 7EJ

Alexander, David
British Red Cross Society Headquarters
9 Grosvenor Crescent
London
SW1X 7EJ

Allan, J A
School of Oriental and African Studies (SOAS)
Thornhaugh Street Russell Square
London
WC1H 0XG

Allen, Penny
Save the Children Fund
Mary Datchelor House 17 Grove Lane
Camberwell
London
SE5 8RD

Allen, Tim
South Bank University
103 Borough Road
London
SE1 0AA

Alley, Eric
Independent Consultant
Farthings Gembling Driffield
North Humberside
YO25 8HS

Allsopp, Andrew
Wind Engineering Society
c/o Institution of Civil Engineers
1 Great George Street
London SW1P 3AA

Alsop, Allen
Ove Arup Partnership
13 Fitzroy Street
London
W1P 6BQ

Ambraseys, N N
Department of Civil Engineering
Imperial College South Kensington
London
SW7 2BU

Anderson, Ewan
Department of Geography
University of Durham South Road
Durham
DH1 1QT

Anderson, Lloyd
Institute of Terrestrial Ecology
Bangor Research Unit Orton Building Deiniol Road
Bangor Gwynedd
LL57 2UP

Arthur, John
Adventist Development and Relief Agency (ADRA)
119 St Peter's Street
St Albans Herts
AL1 3EY

Attwood, Tom
Cargil Attwood Consultants
8 Teddington Park
Teddington
TW11 8DA

Austin, Sally
Care International (UK)
36-38 Southampton Street
London
WC2E 7AF

Bailey, Ron
British Association for Immediate Care
Bay 6-7 Black Horse Lane
Ipswich Suffolk
IP1 2EF

Baines, Jacqueline
UK National Co-ordination Committee for the IDNDR
The Royal Academy of Engineering
29 Great Peter Street
London SW1P 3LW

Baker, Colin
Action Aid
Hamlyn House MacDonald Road Archway
London
N19 5PG

Baker, Colin
Action Aid
Hamlyn House MacDonald Road Archway
London
N19 5PG

Balderstone, David
Adventist Development and Relief Agency (ADRA)
119 St Peter's Street
St Albans Herts
AL1 3EY

Barakat, Sultan
Post War Reconstruction & Development Unit (PRDU)
IoAAS University of York The King's Manor
York YO1 2EP

Barber, Susan
World Vision UK
599 Avebury Boulevard Central
Milton Keynes
MK9 3PG

Barnes, Ian
World Conservation Monitoring Centre
219 Huntingdon Road
Cambridge
CB3 0DL

Baskett, Peter
World Association for Disaster and Emergency Medicine
Accident and Emergency Department
Royal London Hospital
London E1 1BB

Battersby, Neil
Action Water Charity
Mount Hawke Truro
Cornwall
TR4 8BZ

Baxter, Maggie
Charity Projects (Comic Relief)
1st Floor 74 New Oxford Street
London
WC1A 1EF

Baxter, Peter
Department of Community Medicine
Fenners Gresham Road
Cambridge
CB1 2ES

Baxter, Ray
ETC (UK)
117 Norfolk Street North Shields
Tyne and Wear
NE30 1NQ

Baysham, Leo
Christian Aid
P. O. Box 100
London
SE1 7 RT

Bell, Lizzie
Care International (UK)
36-38 Southampton Street
London
WC2E 7AF

Bellers, Roger
Oxford Centre for Disaster Studies (OCDS)
PO Box 137
Oxford
OX4 1BB

Belshaw, Derek
Overseas Development Group
School of Development Studies
University of East Anglia
Norwich NR4 7TJ

Benbow, Indira
Teaching Aids at Low Cost (TALC)
P.O. Box 49
St Albans Herts
AL1 4AX

Bennett, Anne
Quaker Peace and Service
Friends' House Euston Road
London
NW1 2BJ

Bennett, Jon
Independent Consultant
84 Sanfield Road
Oxford
OX3 7RL

Benson, Charlotte
Save the Children Fund
Mary Datchelor House
17 Grove Lane Camberwell
London
SE5 8RD

Besse, Christopher
Medical Emergency Relief International (MERLIN)
1a Rede Place Chepstow Place
London
W2 4TU

Bevan, Philippa
Centre for the Study of African Economies
21 Winchester Road
Oxford
OX2 6NA

Beyani, Chalpka
Refugee Studies Programme
Queen Elizabeth House University of Oxford
21 St Giles
Oxford OX1 3LA

Bignell, V
Systems Group Faculty of Technology
The Open University
Walton Hall
Milton Keynes Beds MK7 6AA

Blaikie, Piers
Overseas Development Group
School of Development Studies University of East Anglia
Norwich
NR4 7TJ

Boam, T A
British Consultants Bureau
1 Westminster Palace Gardens
1-7 Artillery Row
London
SW1P 1RJ

Bond, Kevin
National Rivers Authority
Rivers House Waterside Drive Aztec West
Almondsbury
Bristol BS12 4UD

Booth, Ed
Ove Arup Partnership
13 Fitzroy Street
London
W1P 6BQ

Borden, Jenny
Christian Aid
P. O. Box 100
London
SE1 7 RT

Borton, John
Overseas Development Institute
Regent's College Inner Circle
Regent's Park
London NW1 4NS

Boyden, Jo
International NGO Training & Research Centre (INTRAC)
P.O. Box 563
Oxford OX2 6RT

Bradley, Jo
International Extension College
Dale's Brewery Gwydir Street
Cambridge
CB1 2LJ

Bradstock, Alastair
Farm Africa
9-10 Southampton Place Bloomsbury
London
WC1A 2DA

Briottet, Roger
World Development Movement
25 Beehive Place
London
SW9 7QR

Browitt, Chris
British Geological Survey
Murchison House
West Mains Road
Edinburgh EH9 3LA

Buchanan-Smith, Margie
Action Aid
Hamlyn House MacDonald Road Archway
London
N19 5PG

Burton, Peter
ODA (Emergency Aid Department)
Overseas Development Administration
94 Victoria Street
London SW1E 5JL

Cameron, John
School of Development Studies
University of East Anglia
Norwich
NR4 7TJ

Campbell, David
Farm Africa
9-10 Southampton Place Bloomsbury
London
WC1A 2DA

Cannon, Terry
School of Humanities
University of Greenwich
Wellington Street Greenwich
London
SE18 6PF

Chantler, J
Pell Frischmann Consulting Engineers Ltd.
5 Manchester Square
London
W1A 1AU

Cherrett, Ian
ETC (UK)
117 Norfolk Street North Shields
Tyne and Wear
NE30 1NQ

Chester, David
Department of Geography
University of Liverpool
Liverpool
L69 3BX

Chingdno, Mark
Refugee Studies Programme
Queen Elizabeth House University of Oxford
21 St Giles
Oxford
OX1 3LA

Chmoulian, A
Pell Frischmann Consulting Engineers Ltd.
5 Manchester Square
London
W1A 1AU

Chryssanthopoulos, M
Depart. of Civil Engineering
Imperial College South Kensington
London
SW7 2BU

Clark, Andrew C
Quaker Peace and Service
Friends' House Euston Road
London
NW1 2BJ

Coburn, Andrew
Cambridge Architectural Research Ltd
The Eden Centre
47 City Road
Cambridge
CB1 1DP

Coburn, Andy
Department of Community Medicine
Fenners Gresham Road
Cambridge
CB1 2ES

Cockburn, Charles
Post War Reconstruction & Development Unit (PRDU)
IoAAS University of York The King's Manor
York
YO1 2EP

Cole, Tim
Christian Aid
P. O. Box 100
London
SE1 7 RT

Collins, Mark
World Conservation Monitoring Centre
219 Huntingdon Road
Cambridge
CB3 0DL

Conway, D
Climatic Research Unit School of Environmental Science
University of East Anglia
Norwich
NR4 7TJ

Cooper, Ruth
The Royal Society
6 Carlton House Terrace
London
SW1Y 5AG.

Cotterill, Paul
Action Health
The Gate House 25 Gwydir Street
Cambridge
CB1 2LG

Belinda Cowden
Oxford Centre for Disaster Studies (OCDS)
P O Box 137
Oxford
OX4 1BB

Cowley, Darryl
Karuna Trust
186 Cowley Road
Oxford
OX4 1VE

Crooks, Bill
Tear Fund
100 Church Rd
Teddington Middx
TW11 8QE

Crosthwaite, John
Crown Agents
St Nicholas House St Nicholas Road
Sutton Surrey
SM1 1EL

Culshaw, M
British Geological Survey
Keyworth
Nottingham
NG12 5GG

Cutts, Felicity
London School of Hygiene & Tropical Medicine
Keppel Street
London
WC1E 7HT

Daicon, Diane
Building and Social Housing Foundation
Memorial Square Coalville
Leicestershire
LE67 3TU

Dalton, Mark
Medical Emergency Relief International (MERLIN)
1a Rede Place Chepstow Place
London
W2 4TU

Dalzel, Howard
Concern Worldwide
248-250 Lavender Hill
London
SW11 1LJ

Darby, Eileen
Centre for International Health
University of Wales College of Medicine
Heath Park
Cardiff
CF4 4XN

Davies, Susanna
Institute of Development Studies
University of Sussex Falmer
Brighton
BN1 9RE

Davis, Ian
Oxford Centre for Disaster Studies (OCDS)
PO Box 137
Oxford
OX4 1BB

Degg, Martin
Department of Geography
Chester College Cheyney Road
Chester
GH1 4BJ

Denham, Keith
Llewelyn Davies Planning
Brook House 2-16 Torrington Place
London
WC1E 7HN

Dico, Tess
Y Care International
640 Forest Road
London
E17 3DZ

Diskett, Patricia
Oxfam (Emergency Department)
274 Banbury Road
Oxford
OX2 7DZ

Dodds, Tony
International Extension College
Dale's Brewery Gwydir Street
Cambridge
CB1 2LJ

Douglas, Janet
ODA (Emergency Aid Department)
Overseas Development Administration
94 Victoria Street
London SW1E 5JL

Downing, Tom
Environmental Change Unit
1A Mansfield Road
Oxford
OX1 3TB

Rosenberg, Jane
Natural Resources Institute
Resource Management Division Central Avenue
Chatham Maritime Chatham
Kent
ME4 4TB

Dudley, Eric
Cambridge Architectural Research Ltd
The Eden Centre 47 City Road
Cambridge
CB1 1DP

Duffield, Mark
Centre for Urban & Regional Studies School of Public Policy
JG Smith Building University of Birmingham
Edgbaston
Birmingham B15 2TT

Duncan, L
Centre for Developing Areas Research (CEDAR)
Department of Geography
University of London Egham
London TW20 0EX

Edan, Michael
Centre for Developing Areas Research (CEDAR)
Department of Geography Royal Holloway
University of London Egham
London
TW20 0EX

Ede, David
RED R (Registered Engineers for Disaster Relief)
1-7 Great George Street
London
SW1P 3AA

Eele, Graham
Food Studies Group
Queen Elizabeth House University of Oxford
21 St Giles
Oxford
OX1 3LA

Ellinas, C P
Mott MacDonald Group
St Anne House
20/26 Wellesley House
Croyden Surrey
CR9 2UL

Ellis, Sue
The Research Centre
University of Luton 24 Crawley Green Road
Luton Beds
LU1 3LF

Ellis-Jones, Jim
Silsoe Research Institute
West Park Silsoe
Bedford
MK45 4HS

Elnashai, A S
Department of Civil Engineering
Imperial College South Kensington
London
SW7 2BU

Evans, Brian
Earth Resources Centre
University of Exeter Laver Building
North Park Road
Exeter
EX4 4QE

Fennell, James
Care International (UK)
36-38 Southampton Street
London
WC2E 7AF

Finucane, Jack
Concern Worldwide
248-250 Lavender Hill
London
SW11 1LJ

Fisher, Judith
World Association for Disaster and Emergency Medicine
Accident and Emergency Dept Royal London Hospital Whitechapel
London E1 1BB

Fisher, Simon
Responding To Conflict
Woodrooke College 1046 Bristol Road
Birmingham
B29 6LJ

Flaten, Wynn
Afghanaid
292 Pentonville Road
London
N1 9NR

Fordham, Maureen
Flood Hazard Research Centre
Middlesex University Queensway
Enfield Middx
EN3 4SF

Fortune, J
Systems Group Faculty of Technology
The Open University
Walton Hall
Milton Keynes
MK7 6AA

Fulton, Duncan
SOS Sahel
1 Tolpuddle Street
London
N1 0XT

Garforth, Chris
Agricultural Extension & Rural Development Dept
The University of Reading 3 Earley Gate
P. O. Box 238 The University
Reading RG6 6AL

George, David
Ove Arup Partnership
13 Fitzroy Street
London
W1P 6BQ

Gibbins, Maureen B
Institute of Risk Management
Lloyds Avenue House
6 Lloyd's Avenue
London
EC3N 3AX

Gibbs, Sara
International NGO Training and Research Centre (INTRAC)
P.O. Box 563
Oxford
OX2 6RT

Golden, Barbara
Health & Nutritional Status Advisory Unit (HANSA)
Rowett Research Institute
Greenburn Road Bucksburn
Aberdeen Scotland
AB2 9SB

Golden, M H N
Health & Nutritional Status Advisory Unit (HANSA)
Rowett Research Institute
Greenburn Road Bucksburn
Aberdeen Scotland
AB2 9SB

Goodhand, Mike
British Red Cross Society Headquarters
9 Grosvenor Crescent
London
SW1X 7EJ

Gostelow, Paul
British Geological Survey
Keyworth
Nottingham
NG12 5GG

Graham, Kate
Action Health
The Gate House 25 Gwydir Street
Cambridge
CB1 2LG

Grainger, Peter
Earth Resources Centre
University of Exeter Laver Building North Park Road
Exeter
EX4 4QE

Grant, Anne
Green Cross UK
Mayfield Centre 94 West Hill
London
SW15 2UH

Gray, Alan
Entec Europe Ltd
Northumbria House Regent Centre Gosforth
Newcastle upon Tyne
NE3 3PX

Green, Carolyn
ECHO International Health Service
Ullswater Crescent Coulsdon
Surrey
CR5 2HR

Green, Colin
Flood Hazard Research Centre
Middlesex University Queensway
Enfield Middx
EN3 4SF

Green, Reg
Institute of Development Studies
University of Sussex Falmer
Brighton
BN1 9RE

Guerrero, Carlos
Homeless International
Guildford House
20 Queens Road
Coventry
CV1 3EG

Gurney, Charlotte
Earth Observation Sciences Ltd
Broadmede Farnham Business Park
Farnham Surrey
GU9 8QL

Guttmann, Nick
Concern Worldwide
248-250 Lavender Hill
London
SW11 1LJ

Hailey, John
Cranfield School of Management
Cranfield University
Bedford
MK43 0AL

Hall, Nicholas
Intermediate Technology
Myson House Railway Terrace
Rugby
CV21 3HT

Hamblin, Patrick
Sedgwick Global
Bristol and West House
173 Friar Street
Reading
RG1 1BP

Hamdi, Nabeel
Centre for Development and Emergency Planning
(CENDEP)
Oxford Brookes University Gipsy Lane Headington
Oxford
OX3 0BP

Hammond, Roger
Living Earth
Warwick House 106 Harrow Road
London
W2 1XD

Hardiment, Tony
Catholic Fund for Overseas Development (CAFOD)
2 Romero Close Stockwell Road
London
SW9 7YUK

Harper, Malcolm
United Nations Association (UNA)
3 Whitehall Court
London
SW1A 2EL

Harrell-Bond, Barbara
Refugee Studies Programme
Queen Elizabeth House University of Oxford
21 St Giles
Oxford
OX1 3LA

Harris, Ansel
United Kingdom Jewish Aid
33 Seymour Place
London
W1H 6AT

Harris, Michael
United Kingdom Jewish Aid
33 Seymour Place
London
W1H 6AT

Harvey, Anthony
Harvest Help
3-4 Old Bakery Row
Wellington Telford
TF1 1PS

Hay, Roger
Food Studies Group
Queen Elizabeth House University of Oxford
21 St Giles
Oxford
OX1 3LA

Hayes, Alan
RED R (Registered Engineers for Disaster Relief)
1-7 Great George Street
London
SW1P 3AA

Headington, Simon
Health Projects Abroad
P.O. Box 24
Bakewell Derbyshire
DE45 1ZW

Henry, Dr J
Oxford Brookes University
Gipsy Lane Headington
Oxford OX3 OBP

Henry, James
Independent Consultant
Hillcrest Chalk Lane Hyde Heath
Amersham Bucks
HP6 5SA

Hewitt, Barry
Sedgwick UK Ltd
Bristol and West House 173 Friar Street
Reading
RG1 1BP

Hillman , P F
Gifford and Partners
Carlton House Ringwood Road Woodlands
Southampton
SO40 7 HT

Hindmarsh, Patricia
Marie Stopes International
62 Grafton Way
London
W1P 5LE

Hindmarsh, Paul
Natural Resources Institute
Resource Management Division Chatham Maritime
Chatham
Kent
ME4 4TB

Hines, K
British Association for Immediate Care
Bay 6-7 Black Horse Lane
Ipswich Suffolk
IP1 2EF

Hinings, Nigel
The Society for Earthquake & Civil Eng. Dynamics (SECED)
1-7 Great Peter Street
London
SW1P 3AA

Hodgkinson, Peter
The Centre for Crisis Psychology
Four Arches Broughton Hall Skipton
Skipton North Yorks
BD23 3AE

Hodgson, Robert L P
Devon Aid
Lower Beer Uplowman
Tiverton Devon
EX16 7PF

Hoffman, Michael
Homeless International
Guildford House
20 Queens Road
Coventry
CV1 3EG

Hogan, Mark
Concern Worldwide
248-250 Lavender Hill
London
SW11 1LJ

Holden, Pat
Multilateral Research Economics Department
ODA
94 Victoria Street
London
SW1E 5JL

Holt, Julius
Save the Children Fund
Mary Datchelor House 17 Grove Lane Camberwell
London
SE5 8RD

Homan, Jacqueline
Department of Geography
Chester College Cheyney Road
Chester
GH1 4BJ

Horlick-Jones, Tom
Centre for Environmental Strategy
University of Surrey
Guildford
Surrey GU2 5HX

Horning, M
Institute of Terrestrial Ecology
Bangor Research Unit Orton Building
Deiniol Road
Bangor Gwynedd
LL57 2UP

Horsnell, Julia
ODA (Emergency Aid Department)
Overseas Development Administration 94 Victoria Street
London
SW1E 5JL

Howell, Philippa
Action Aid
Hamlyn House MacDonald Road Archway
London
N19 5PG

Hughes, A
International Seismological Centre
Piper's Lane Thatcham
Newbury
RG13 4NS

Hulme, David
Institute for Development Policy and Management (IDPM)
University of Manchester Precinct Centre
Oxford Rd
Manchester M13 9GH

Hulme, Mike
Climatic Research Unit School of Environmental Science
University of East Anglia
Norwich
NR4 7TJ

Hunt, Julian
Hadley Centre Meteorological Office
London Road
Bracknell Berks
RG12 2SZ

Irwin, Michael
United Nations Association (UNA)
3 Whitehall Court
London
SW1 2EL

Jackson, A A
Institute of Civil Defence and Disaster Studies
Bell Court House 11 Bloomfield Street
London
EC2M 7AY

James, Peter
Agency for Research & Co-operation in Development (ACORD)
Francis House Francis Street
London
SW1P 1DQ

James, W P T
Health & Nutritional Status Advisory Unit (HANSA)
Rowett Research Institute Greenburn Road Bucksburn
Aberdeen
AB2 9SB

Jones, David K C
Hazard and Risk Management Studies (HARMS)
c/o Dept. Geography London School of Economics
Houghton Street
London
WC2A 2AE

Jones, Steve
Rural Resources Management Ltd.
PO Box 144 Bourn
Cambridge
CB3 7SL

Joseph, Peter
Karuna Trust
186 Cowley Road
Oxford
OX4 1VE

Judge, Barbara
London School of Hygiene & Tropical Medicine
Keppel Street
London
WC1E 7HT

Kapila, Mukesh
Multilateral Research Economics Department (ODA)
94 Victoria Street
London
SW1E 5JL

Kelly, P M
Climatic Research Unit School of Environmental Science
University of East Anglia
Norwich NR4 7TJ

Kendall, Simon
SAFE (Support Action for Emergencies)
Cumbers Liss
Hampshire
GU33 7LL

Kerr, C
Department of Civil Engineering
Imperial College South Kensington
London
SW7 2BU

Kinsella, Mary
The Society for Earthquake & Civil Eng. Dynamics (SECED)
1-7 Great Peter Street
London
SW1P 3AA

Kirkby, John
ETC (UK)
117 Norfolk Street North Shields
Tyne and Wear
NE30 1NQ

Knill, Sir John
Independent Consultant (Chair, UK IDNDR Committee)
Highwood Farm Shaw-cum-Donnington
Newbury Berkshire
RG16 9LB

Koch de Gooryend, Peter
Care International (UK)
36-38 Southampton Street
London
WC2E 7AF

Koteeswaran, V
Pell Frischmann Consulting Engineers Ltd
5 Manchester Square
London
W1A 1AU

Lambert, Bobby
RED R (Registered Engineers for Disaster Relief)
1-7 Great George Street
London
SW1P 3AA

Large, Judith
Responding To Conflict
Woodrooke College 1046 Bristol Road
Birmingham
B29 6LJ

Lawes, Howard
Noble Denton Weather Services Ltd
Noble House 131 Aldersgate Street
London
EC1A 4EB

LeChene, E
Institute of Civil Defence and Disaster Studies
Bell Court House 11 Bloomfield Street
London
EC2M 7AY

Leddon, David
Overseas Development Group
School of Development Studies University of East Anglia
Norwich
NR4 7TJ

Lee, Brian
Department of Civil Engineering
University of Portsmouth
Portsmouth
PO1 3QL

Leonard, John B
Centre for Environmental and Human Settlements (CEHS)
Edinburgh College of Art Heriot-Watt University
Lauriston Place
Edinburgh EH3 9DF

Lewis, David
Independent Consultant
117 Gray's Inn Buildings Rosebery Ave
London
EC1R 4PN

Lewis, James
Datum International
101 High Street Marshfield Chippenham
Wiltshire
SN6 8LA

Lewis, Terri
International Christian Relief
PO Box 180 16 St Johns Hill Sevenoaks
Kent
TN13 3NP

Logan, Richard
National Rivers Authority
Rivers House Waterside Drive
Aztec West Almondsbury
Bristol
BS12 4UD

Longhurst, Richard
Institute of Child Health
University of London
London
WC1N 1EH

Longworth, M
Hadley Centre Meteorological Office
London Road
Bracknell Berks
RG12 2SZ

Luxmoore, Richard
World Conservation Monitoring Centre
219 Huntingdon Road
Cambridge
CB3 0DL

Lynagh, Norman
Noble Denton Weather Services Ltd
Noble House 131 Aldersgate Street
London
EC1A 4EB

Lyne, W H
Hadley Centre Meteorological Office
London Road
Bracknell Berks
RG12 2SZ

Macrae, Joanna
Overseas Development Institute
Regent's College Inner Circle
Regent's Park
London
NW1 4NS

Magurie, J
Lloyd's Register (Civil and Structural Eng.)
29 Wellesley Road Croyden
Surrey
CR0 2AJ

Manson, Felicity
Y Care International
640 Forest Road
London
E17 3DZ

Mardel, S
UK- Med
North Staffordshire Royal Infirmary
Princes Road Hartshill
Stoke-on-Trent
ST4 7LN

Marriott, Niall
Living Earth
Warwick House 106 Harrow Road
London
W2 1XD

Martin, Chris
South Bank University
103 Borough Road
London
SE1 0AA

Maskrey, Andrew
Intermediate Technology
Myson House Railway Terrace
Rugby
CV21 3HT

Matthews, Alan
Crown Agents
St Nicholas House St Nicholas Road
Sutton Surrey
SM1 1EL

Matthews, J C
British Consultants Bureau
1 Westminster Palace Gardens
1-7 Artillery Row
London
SW1P 1RJ

Mcerlean, Joan
Scottish Catholic International Aid Fund (SCIAF)
5 Oswald Street
Glasgow
G1 4QR

McGregor, D M
International Seismological Centre
Piper's Lane Thatcham
Newbury
RG13 4NS

McK. Holloway, J
Institute of Civil Defence and Disaster Studies
Bell Court House 11 Bloomfield Street
London
EC2M 7AY

McKemey, Kevin
Independent Consultant
14 Andrews Road Earley
Reading Berks
RG6 7PJ

McKemey, Kevin
SAFE (Support Action for Emergencies)
Cumbers Liss
Hampshire
GU33 7LL

McLeod, Ruth
Homeless International
Guildford House
20 Queens Road
Coventry
CV1 3EG

Membrey, David
Book Aid International
39-41 Coldharbour Lane Camberwell
London
SE5 9NR

Merefield, John R
Earth Resources Centre
University of Exeter Laver Building
North Park Road
Exeter
EX4 4QE

Miller, Richard
Catholic Fund for Overseas Development (CAFOD)
2 Romero Close Stockwell Road
London
SW9 TYUK

Mills, C F
Health & Nutritional Status Advisory Unit (HANSA)
Rowett Research Institute Greenburn Road Bucksburn
Aberdeen
AB2 9SB

Montague, Simon
Entec Europe Ltd
Northumbria House Regent Centre Gosforth
Newcastle upon Tyne
NE3 3PX

Moore, G H
Independent Consultant
Authorpe House 36 Horncastle Road
Woodhall Spa Lincs
LN10 6UZ

Morgan, Richard
Sedgwick Global
Bristol and West House 173 Friar Street
Reading
RG1 1BP

Motawee, Sharif
Centre for Environmental and Human Settlements (CEHS)
Edinburgh College of Art Heriot-Watt University
Lauriston Place
Edinburgh Scotland EH3 9DF

Mulhier, Mohammed Omer
Cranfield Disaster Preparedness Centre (CDPC)
Cranfield University RMCS Shrivenham
Swindon Wilts
SN6 8LA

Mumtaz, Babar
Development Planning Unit
9 Endsleigh Gardens
London
WC1H OED

Munro, Paula
Centre for African Studies
University of Cambridge Free School Lane
Cambridge
CB2 3RQ

Musson, Roger
British Geological Survey
Murchison House
West Mains Road
Edinburgh
EH9 3LA

Murray, John
Volcano Geophysics Group
Department of Earth Sciences Open University
Walton Hall
Milton Keynes
MK7 6AA

Myers, Mary
Cranfield Disaster Preparedness Centre (CDPC)
Cranfield University RMCS Shrivenham
Swindon Wilts
SN6 8LA

Nevein, J
UK- Med
North Staffordshire Royal Infirmary
Princes Road Hartshill
Stoke-on -Trent
ST4 7LN

O'Connor, Mrs
British Consultants Bureau
1 Westminster Palace Gardens 1-7 Artillery Row
London
SW1P 1RJ

O'Keefe, Phil
ETC (UK)
117 Norfolk Street North Shields
Tyne and Wear
NE30 1NQ

Oakley, David
Independent Consultant
38 The Street Barney Fakenham
Norfolk
NR21 ONB

Oppenhiemer, Clive
Department of Geography University of Cambridge
Downing Place
Cambridge
CB2 3EN

Palmer-Jones, Richard
Overseas Development Group
School of Development Studies
University of East Anglia
Norwich
NR4 7TJ

Palmer-Jones, Richard
School of Development Studies
University of East Anglia
Norwich
NR4 7TJ

Parker, Dennis
Flood Hazard Research Centre
Middlesex University Queensway
Enfield Middx
EN3 4SF

Pasche, Alain
Green Cross UK
Mayfield Centre 94 West Hill
London
SW15 2UH

Penning-Rowshell, Edmund
Flood Hazard Research Centre
Middlesex University Queensway
Enfield Middx
EN3 4SF

Penny, Steve
51 Cambridge Avenue
New Malden
Surrey
KT3 4LD

Pereira, Julie
Medical Emergency Relief International (MERLIN)
1a Rede Place Chepstow Place
London
W2 4TU

Peters, G
Systems Group Faculty of Technology
The Open University
Walton Hall
Milton Keynes Beds
MK7 6AA

Polaikie, Piers
School of Development Studies
University of East Anglia
Norwich
NR4 7TJ

Pomonis, Antonios
Cambridge Architectural Research Ltd
The Eden Centre
47 City Road
Cambridge
CB1 1DP

Pomonis, Tony
Department of Community Medicine
Fenners Gresham Road
Cambridge
CB1 2ES

Pratt, Brian
International NGO Training and Research Centre (INTRAC)
P.O. Box 563
Oxford
OX2 6RT

Prescott, M
UK- Med
North Staffordshire Royal Infirmary Princes Road
Hartshill
Stoke-on-Trent
ST4 7LN

Priestley, Michael
United Nations Association (UNA)
3 Whitehall Court
London
SW1 2EL

Redmond, A
UK- Med
North Staffordshire Royal Infirmary Princes Road
Hartshill
Stoke-on-Trent
ST4 7LN

Reed, R
Water Engineering and Development Centre (WEDC)
Loughborough University of Technology
Loughborough Leics
LE11 3TU

Reedman, A
British Geological Survey
Keyworth
Nottingham
NG12 5GG

Rees, Rob
Catholic Fund for Overseas Development (CAFOD)
2 Romero Close Stockwell Road
London
SW9 TYUK

Reynolds, John
10 Bron y Nant
Mold
Clwyd
North Wales
CH7 1UX

Richman, Naomi
Institute of Child Health (University of London)
30 Guildford Street
London
WC1N 1EH

Rosenhead, Jonathan
Hazard and Risk Management Studies (HARMS)
c/o Dept. Geography London School of Economics
Houghton Street
London
WC2A 2AE

Rowland, Jon
Llewelyn Davies Planning
Brook House 2-16 Torrington Place
London
WC1E 7HN

Rymer, Hazel
Volcano Geophysics Group
Department of Earth Sciences Open University
Walton Hall
Milton Keynes
MK7 6AA

Safier, Michael
Development Planning Unit
9 Endsleigh Gardens
London
WC1H OED

Sanderson, David
Oxford Centre for Disaster Studies (OCDS)
PO Box 137
Oxford
OX4 1BB

Sandford, Stephen
Farm Africa
9-10 Southampton Place Bloomsbury
London
WC1A 2DA

Sanyasi, Anthea
Refugee Studies Programme
Queen Elizabeth House University of Oxford 21 St Giles
Oxford
OX1 3LA

Satterthwaite, David
International Institute for Environment and Development (IIED)
3 Endsleigh Street
London
WC1H 0DD

Savage, P E A
World Association for Disaster and Emergency Medicine
Accident and Emergency Department
Royal London Hospital Whitechapel
London E1 1BB

Scarlett, T
United Kingdom Jewish Aid
33 Seymour Place
London
W1H 6AT

Schilderman, Theo
Intermediate Technology
Myson House Railway Terrace
Rugby
CV21 3HT

Schubert, Michael
Medical Emergency Relief International (MERLIN)
1a Rede Place Chepstow Place
London
W2 4TU

Seaman, John
Save The Children Fund UK
17 Grove Lane Camberwell
London
SE5 8RD

Seidel, Gill
Independent Consultant
c/o Africa Research Unit University of Bradford
Bradford
BD7 1DP

Seidl, Harold
Adventist Development and Relief Agency (ADRA)
119 St Peter's Street
St Albans Herts
AL1 3EY

Shepherd, Andrew
Centre for Urban & Regional Studies
School of Public Policy
JG Smith Building University of Birmingham Edgbaston
Birmingham B15 2TT

Simon, David
Centre for Developing Areas Research (CEDAR)
Department of Geography Royal Holloway
University of London Egham
London
TW20 0EX

Sims, Brian
Silsoe Research Institute
Wrest Park Silsoe
Bedford
MK45 4HS

Sinclair, Douglas
Adventist Development and Relief Agency (ADRA)
119 St Peter's Street
St Albans Herts
AL1 3EY

Slatter, Keith
ECHO International Health Service
Ullswater Crescent Coulsdon
Surrey
CR5 2HR

Slim, Hugo
Centre for Development and Emergency Planning (CENDEP)
Oxford Brookes University Gipsy Lane
Headington
Oxford OX3 0BP

Smith, M
Water Engineering and Development Centre (WEDC)
Loughborough University of Technology
Loughborough Leics
LE11 3TU

Smith, P
Centre for Arid Zone Studies
Thoday Building University of Wales Bangor
Gwynedd
LL57 2UW

Smith, Robert
UK Committee for UNICEF
55 Lincoln Inn Fields
London
WC2 3NB

Spence, Robin
Department of Community Medicine
Fenners Gresham Road
Cambridge
CB1 2ES

Staines, B
Institute of Terrestrial Ecology
Bangor Research Unit Orton Building Deiniol Road
Bangor Gwynedd
LL57 2UP

Stanton, Patrick
Institute of Civil Defence and Disaster Studies
Bell Court House 11 Bloomfield Street
London
EC2M 7AY

Stephenson, Robin
Independent Consultant
22 Sunderland Road Ealing
London
W5 4JY

Stewart, Michael
The Centre for Crisis Psychology
Four Arches Broughton Hall Skipton
Skipton North Yorks
BD23 3AE

Stockton, Nicholas
Oxfam (Emergency Department)
274 Banbury Road
Oxford
OX2 7DZ

Stordy, Stephen
Y Care International
640 Forest Road
London
E17 3DZ

Stuttard, Matthew
Earth Observation Sciences Ltd
Broadmede Farnham Business Park
Farnham Surrey
GU9 8QL

Sweeney, Mary
Trocaire
169 Booterestoon Avenue Blackrock
Dublin Ireland

Swift, Jeremy
Institute of Development Studies
University of Sussex Falmer
Brighton
BN1 9RE

Taylor, C A
Earthquake Engineering Research Centre
Department of Civil Engineering University Walk
Bristol University
Bristol
BS8 1TR

Taylor, Jo
World Conservation Monitoring Centre
219 Huntingdon Road
Cambridge
CB3 0DL

Templer, Guy
Multilateral Research Economics Department
ODA
94 Victoria Street
London
SW1E 5JL

Thomas, John
International Extension College
Dale's Brewery Gwydir Street
Cambridge
CB1 2LJ

Thompson, Donald
Centre for Developing Areas Research (CEDAR)
Department of Geography Royal Holloway
University of London Egham
London
TW20 0EX

Thompson, Marcus
Oxfam (Emergency Department)
274 Banbury Road
Oxford
OX2 7DZ

Thorsen, Leigh
Y Care International
640 Forest Road
London
E17 3DZ

Toibin, Niall
Trocaire
169 Booterestoon Avenue Blackrock
Dublin Ireland

Toomer, C
Sedgwick Global
Bristol and West House 173 Friar Street
Reading
RG1 1BP

Tricklebank, Alan
Gifford and Partners
Carlton House Ringwood Road Woodlands
Southampton
SO40 7 HT

Turbitt, Terry
British Geological Survey
Murchison House
West Mains Road
Edinburgh
EH9 3LA

Turton, David
Department of Social Anthropology
University of Manchester Roscoe Building 5th Floor
Brunswick Street
Manchester
M13 9PLUK

Twigg, John
Intermediate Technology
Myson House Railway Terrace
Rugby
CV21 3HT

Van den Hurk, Rudi
International Christian Relief
PO Box 180 16 St Johns Hill Sevenoaks
Kent
TN13 3NP

van Oosten, Hans
PLAN International
Chobham House Christchurch Way
Woking Surrey
GU21 1JG

Vaux, Tony
Oxfam (Emergency Department)
274 Banbury Road
Oxford
OX2 7DZ

Vaughan, Gwen
Living Earth
Warwick House 106 Harrow Road
London
W2 1XD

Vaughan, Hilary
Crown Agents
St Nicholas House St Nicholas Road
Sutton Surrey
SM1 1EL

Venables, Hugh
Saferworld
3rd Floor 34 Alfred Place
London
WC1E 7DP

Village, Graham
Catastrophe Reinsurance
DYP Group Limited Bridge House
181 Queen Victoria Street
London
EC4V 4DD

Wadge, Geoff
NERC Unit for Thematic Information Systems
Department of Geology University of Reading
Reading
RG6 2AB

Wall, Michael
Tear Fund
100 Church Rd
Teddington Middx
TW11 8QE

Walker, Alice
British Geological Survey
Murchison House
West Mains Road
Edinburgh
EH9 3LA

Walter, Derek
World Aware
1 Catton Street
London
WC1R 4AB

Walton, David
Llewelyn Davies Planning
Brook House 2-16 Torrington Place
London
WC1E 7HN

Ward, Gavin
Post War Reconstruction & Development Unit (PRDU)
IoAAS University of York The King's Manor
York
YO1 2EP

Wells, Victoria
ECHO International Health Service
Ullswater Crescent Coulsdon
Surrey
CR5 2HR

Welsh, Liz
Centre for the Study of African Economies
21 Winchester Road
Oxford
OX2 6NA

Westgate, Ken
Cranfield Disaster Preparedness Centre (CDPC)
Cranfield University RMCS Shrivenham
Swindon Wilts
SN6 8LA

White, David
Harvest Help
3-4 Old Bakery Row
Wellington Telford
TF1 1PS

Wilcocks, Theo
Silsoe Research Institute
Wrest Park Silsoe
Bedford
MK45 4HS

Wilkinson, Peter
World Vision UK
599 Avebury Boulevard Central
Milton Keynes Bucks
MK9 3PG

Williams, Rik
World Vision UK
599 Avebury Boulevard Central
Milton Keynes Bucks
MK9 3PG

Wolf, Peter
Pell Frischmann Consulting Engineers Ltd.
5 Manchester Square
London
W1A 1AU

Wright, Colin E
Department of the Environment
Room A4.32 Romney House
43 Marsham St
London
SW1P 3PY

Wright, D
Centre for Arid Zone Studies
Thoday Building University of Wales Bangor
Gwynedd
LL57 2UW

Wyer, June
Christian Aid
P. O. Box 100
London
SE1 7 RT

Yoshida, Hito
School of Oriental and African Studies (SOAS)
Thornhaugh Street Russell Square
London
WC1H 0XG

Zwi, Anthony
London School of Hygiene & Tropical Medicine
Keppel Street
London
WC1E 7HT

Appendices

One
World Conference on Natural Disaster Reduction, Yokohama, 1994; Strategy for the Year 2000 and Beyond

Two
Current networks

Three
Methodology summary
The audit questionnaire

Appendix one

World Conference on Natural Disaster Reduction, Yokohama, 1994; Strategy for the Year 2000 and Beyond

The World Conference, based on adoption of the principles and the assessment of the progress accomplished during the first half of the Decade, has formulated a strategy for disaster reduction centred on the objective of saving human lives and protecting property. The strategy calls for an accelerated implementation of a Plan of Action to be developed from the following points:

A. Development of a global culture of prevention as an essential component of an integrated approach to disaster reduction.

B. Adoption of a policy of self-reliance in each vulnerable country and community comprising capacity-building as well as allocation and efficient use of resources.

C. Education and training in disaster prevention, preparedness and mitigation.

D. Development and strengthening of human resources and material capabilities and capacity of research and development institutions for disaster reduction and mitigation.

E. Identification and networking of existing centres of excellence so as to enhance disaster prevention, reduction and mitigation activities.

F. Improvement of awareness in vulnerable communities, through a more active and constructive role of the media in respect of disaster reduction.

G. Involvement and active participation of the people in disaster reduction, prevention and preparedness, leading to improved risk management.

H. In the second half of the decade, emphasis should be given to programmes that promote community-based approaches to vulnerability reduction.

I. Improved risk assessment, broader monitoring and communication of forecasts and warnings.

J. Adoption of integrated policies for prevention of, preparedness for, and response to, natural disasters and other disaster situations including environmental and technological hazards.

K. Improved co-ordination and co-operation among ongoing national, regional and international disaster research activities, at universities, regional and subregional organisations and other technical and scientific institutions, having in mind that links between causes and effects, inherent to all types of disaster, should be investigated through interdisciplinary research.

L. Effective national legislation and administrative action, higher priority at the political decision-making level.

M. Placing higher priority on the compilation and exchange of information on natural disaster reduction, especially at regional and subregional levels, through the strengthening of existing mechanisms and improved use of communications techniques.

N. Promotion of regional and sub-regional co-operation between countries exposed to the same natural hazards through exchange of information, joint disaster reduction activities and other formal or informal means including the establishment or strengthening of regional and sub-regional centres.

O. Making available the existing technology for broader application to disaster reduction.

P. Integration of the private sector in disaster reduction efforts through promotion of business opportunities.

Q. Promotion of the involvement of non-government organisations in natural hazard management, in particular those dealing with environmental and related issues and including indigenous non-governmental organisations,

R. Strengthening the capacity of the United Nations system to assist in the reduction of losses from natural and related technological disasters, including co-ordination and evaluation of activities through the decade and other mechanisms.

Appendix two
Current networks

This is a list of all the entries given together with the names of those who are members. The entries are those that were sent in by respondents, without alteration except to correct misspellings or make consistent entries for the same organisation. The list therefore includes overseas networks and electronic networks together with a number of miscellaneous organisations. Members are listed in brackets with each entry.

ACBAR
ACC/SCN Refugee Nutrition Information System
ACC/SCN Acting Committee on Co-ordination
ACP
AES
Association of Geoscientists in Development (Martin Degg/Dept of Geography Chester College)
Association of Consulting Engineers (Simon Montague/Entec Europe Ltd)
Association of MBAs (Robin Stephenson)
Bangladesh Disaster Forum, Dhaka
BCB (Alan Mathews/Crown Agents)
Bellagio Publishing Network
BOND (John McCall/International Extension College; Simon Headington/Health Projects Abroad)
British Agencies Afghanistan Group
British Consultants Bureau (Simon Montague/Entec Europe Ltd)
British Dam Society
British Hydrological Society
British Invisibles (Alan Mathews/Crown Agents)
British Nutrition Society
British Psychological Society (PE Hodgkinson/Centre for Crisis Psychology)
Caritas International (Joan Mcerlean/Scottish Catholic International Aid Fund)
Centre for Development Studies, Bath University
CIDSE (Joan Mcerlean/Scottish Catholic International Aid Fund)
CLONG
CODEP (Maggie Baxter/Charity Projects; John Potter/London School of Hygiene and Tropical Medicine)
Commission and Association within Meteorological (WH Lyne/Hadley Centre Meteorological Office)
Co-ordinating Committee for Conflict Resolution
Council of Churches of Britain and Ireland

Development Studies Association (Robin Stephenson; John Potter/London School of Hygiene and Tropical Medicine)
Disasters Emergency Committee (Chris Beer/HelpAge International)
Earthquake Engineering Field Investigation Team
Earthquake Engineering Research Institute, USA
EC-NGO network
EMA Rapid Response to Major Disasters
Emergency Planning Society (Eric Alley)
EuronAid
European Association for Studies of Science and Technology
European Network of Bangladesh Studies
Evangelical Missionary Alliance
Forced Migration Network, University of Oxford (John B Leonard/Edinburgh College)
Geological Society London
Global Seismic Data Network
Hazard Network, University of Arizona (John B Leonard/Edinburgh College)
Hazard Forum
HIC (Ruth McLeod/Homeless International)
Huridocs
IAAE
IBRASAM
ICD and DS
IMAGE (John McCall/International Extension College)
INCS Forum (NGOs with interest in Somalia) (Maggie Baxter/Charity Projects)
Informal Social Studies of Visti Community
Institute of British Geographers
Institute of Management Consultants
Institution of Civil Defence and Disaster Studies
Institution of Civil Engineers (BE Lee/Dept of Civil Engineering University of Portsmouth)
Institute of Risk Management
Interchurch Relief and Development Alliance
International Association of Wind Engineering
International Chamber of Commerce (Alan Mathews/Crown Agents)
International Civil Defence Organisation (Geneva)
International Council for Protection of Monuments
International Geographical Union Famine Commission
International Society of Travel Medicine (AD Redmond/UK-Med)
IPCC (M Hulme/Climatic Research Unit University of East Anglia)
IRC Water News
IRDN
ITDG
Joint Association for Wind Engineering (BE Lee/Dept of Civil Engineering University of Portsmouth)

London School of Hygiene and Tropical Medicine (Christopher Bes/MERLIN)
ODI/ODI Network
Pastoral Development Network (ODI)
Plunkett Foundation
PRDU York University
PW Worldwide
RAPID ED (network of NGOs in education) (John McCall/International Extension College)
Refugee Participation Network (John Potter/London School of Hygiene and Tropical Medicine)
Registered Engineers for Disaster Relief (RedR)
Relief and Rehabilitation Network (David Morley/Institute of Child Health, Susan Barber/World Vision)
RIBA
Risk and Reliability Society (CP Ellinas/Mott MacDonald Group)
Royal Academy of Engineering
Royal Meteorological Society (M Hulme/Climatic Research Unit University of E Anglia)
Royal United Services Institute (RUSI)
RTPI
SID
Silver Platter/Medicine (J Fisher/World Association for Disaster)
Society of Industrial Emergency Services Officers
Society for Risk Analysis
Society of Risk Management
Structural Adjustment Forum
Sudan Lobby Group (Maggie Baxter/Charity Projects)
TAA
TAMS (Maureen B Gibbins/Institute of Risk Management)
Technology Management in Europe
UDG
UK Clinical and Public Health Medical Organisation (P Baxter/Dept of Community Medicine Cambridge University)
UK/EU Advisory European EUAC Society
UK IDNDR committees/working groups (David Oakley)
UK NGO AIDS Consortium
UK Systems Society
UK Working Group on Landmines (Maggie Baxter/Charity Projects)
UN Reform Group, London
UNHCR Information Notes on Former Yugoslavia
Volcanic Studies Group (Geological Society)
Volcano Net/Network (P Baxter/Dept of Community Medicine University of Cambridge; RSJ SP/Geology Dept University of Bristol)
Wind Engineering Society (BE Lee/Dept of Civil Engineering University of Portsmouth)
World Association of Emergency and Disaster Medicine (AD Redmond/UK-Med)
World Health Organisation Disaster Assessor, Europe (P Baxter/Dept of Community Medicine University of Cambridge)
World Health Organisation/UNHCR Information Service

Other networks known to be operating in the UK

International Association for the Study of Forced Migration
Royal Institute for International Affairs
Chatham House
10 St James's Square
London SW1Y 4LE

Professional

Association of Geoscientists in Development
Association of Consulting Engineers
Association of MBAs
Bellagio Publishing Network
British Consultants Bureau
British Dam Society
British Hydrological Society
British Nutrition Society
British Psychological Society
Commission and Association within
Council of Churches of Britain and Ireland
Evangelical Missionary Alliance
Geological Society London
Institute/Institution of Civil Engineers
Institute of British Geographers
Institute of Management Consultants
Joint Association for Wind Engineering
London School of Hygiene and Tropical Medicine
Royal Academy of Engineering
RIBA
Royal Meteorological Society
Silver Platter/Medicine
Society for Risk Analysis
UK Clinical and Public Health Medical Organisation
UK Systems Society
Wind Engineering Society
International Society of Travel Medicine
International Chamber of Commerce
International Association of Wind Engineering

General development
BOND
Centre for Development Studies (Bath U)
DSA
EC-NGO network
Pastoral Development Network (ODI)
Structural Adjustment Forum
UN Reform Group, London

Other (& unknown)

AES
ACP
BCB

CLONG
European Association for Studies of Science and Technology
HIC
Huridocs
IAAE
IBRASAM
ICD and DS
IMAGE
Informal Social Studies of Visti Community
International Council for Protection of Monuments
International Panel on Climate Change
IRC Water News
IRDN
ITDG
ODI/ODI Network
Plunkett Foundation
PW Worldwide
RAPID ED (network of NGOs in education)
Royal United Services Institute (RUSI)
RTPI
SID
TAA
TAMS
Technology Management in Europe
UDG
UK/EU Advisory European EUAC Society
UK Systems Society

list of networks by type

Disaster (UK)
Emergency Planning Society
CODEP (UK Network on Conflict, Development and Peace)
Co-ordinating Committee for Conflict Resolution
DEC
EMA Rapid Response to Major Disasters
Forced Migration Network, University of Oxford
Hazard Forum
Institution of Civil Defence and Disaster Studies
Institute of Risk Management
Interchurch Relief and Development Alliance
PRDU York University
Refugee Participation Network
Registered Engineers for Disaster Relief (RedR)
Relief and Rehabilitation Network (ODI)
Risk and Reliability Society
Society of Risk Management
UK IDNDR committees/working groups
UK NGO AIDS Consortium
UK Working Group on Landmines
Volcanic Studies Group (Geological Society)
Volcano NET

International
ACC/SCN Refugee Nutrition Information System
ACC/SCN Acting Committee on Co-ordination
Earthquake Engineering Field Investigation Team - [?USA]
Earthquake Engineering Research Institute, USA
Global Seismic Data Network
Hazard Network, University of Arizona
International Civil Defence Organisation (Geneva)
International Geographical Union Famine Commission
Society of Industrial Emergency Services Officers
UNHCR Information Notes on Former Yugoslavia
World Association of Emergency and Disaster Medicine
WHO Disaster Assessor (Europe)
WHO/UNHCR Information Service

Geographical
ACBAR
Bangladesh Disaster Forum
British Agencies Afghanistan Group
European Network of Bangladesh Studies
INCS Forum (NGOs with interest in Somalia)
Sudan Lobby Group

Appendix three
Summary of research methodology

The aim of the audit was to gain a coherent overview of the current UK assets available in the field of Disaster Mitigation and Preparedness (DMP), in order to:

- Identify strengths, weaknesses and gaps in UK DMP capability;
- Encourage and enhance networking and skillsharing within the DMP community;
- Contribute to a more comprehensive and co-ordinated response to international disaster needs.

At the outset of the project the intention had been for a series of face to face interviews of all those working in disaster preparedness and mitigation. However it soon became apparent that the DMP community was far larger than originally suspected. Hence a postal questionnaire was developed (see below).

The questionnaire was divided into four sections:

- *Organisations* (general information, ie name, address, etc);
- *Activities* (information on regions of activity, hazard expertise, skills available, work content, education and training courses, professional networks and funding);
- *Information* (for example, journals, information and networks);
- *Courses* (training and academic courses available within the UK).

Over 450 questionnaires were mailed, using addresses gained from existing databases of NGO and disaster organisations (including OCDS's), as well as networks, eg CODEP. The strategy of mailing was very much to mailshot as many organisations and individuals as possible, in order to try not to miss any organisations and/or individuals who were not known to the 'mainstream' DMP community.

To complement the questionnaires selected representational organisations and individuals were interviewed, the findings forming the basis of the caption boxes. The aim of the boxes is to present a more qualitative feel for the information presented.

Findings

At the end of the exercise 173 completed questionnaires were returned. This figure is low in comparison to the 450 sent out, yet encouraging when the strategy for mailing, as stated, was that if an organisation was in the remotest way suspected of working in this field, then a questionnaire was sent. The *proportional* percentage of returned questionnaires according to organisational type was:

Organisational type	Proportional per cent
Charity/NGO	33
Individual consultant	11
Private company	7
Government department	2
Academic/research body	31
Intergovernmental agency	1
Consultancy	15

Therefore, for example, it can be seen that one third of all returning questionnaires were from charities/NGOs.

The database

A key component of this audit was the development of a database of information. The database contains all the information represented in this publication. It contains each of the main sections of the audit and allows for cross interrogation of the following categories: organisations, regions of activity, hazard expertise, work content and skills. There were no restrictions on information entered into the database, other than that the information followed the key theme of disaster mitigation, preparedness and (eventually widened to) response.

The database software used was *Microsoft Access Two*. The database structure allowed for questioning of the data, as well as flexibility in presentation. It should be noted therefore that this report represents only an overview of the data available, since the interrogations that can be made of the data are almost infinite. Findings were analysed both quantitatively (the statistics as presented in the tables and charts) and qualitatively - the editors matched their own experience and knowledge, as well as did the first draft readers from IDNDR Committees) with the data being produced.

Data presentation

A decision was made to structure the findings primarily according to the constant of *organisation* relative to respective organisations' *activities* (including regions of activity, hazard expertise, work content and skills). Hence key chapters are presented by activities. For each of these sections there are two sets of statistics presented: a pie chart giving the breakdown of organisations according to activity, for example 8 per cent of all organisations responding to the audit are involved in landslide related activities; and a table giving the percentage of activity of a particular organisation according to activity, for example 37 per cent of individual consultants replying to the questionnaires stated work in famine.

As stated in the foreword, it is important to make the following general points regarding the audit and its interpretation: the responses to the questionnaires and to individual interviews was encouraging. However, there are some gaps: some individuals and organisations failed to submit their response in time for inclusion in the analysis (although they are included in the directory). The editors may not also have been able to contact all who should have been included in this exercise. It is hoped that later editions will cover such omissions. The information contained in this audit reflects the responses given on the returned questionnaires: it was felt *not* to be the role of the editors to alter any returning information (unless obviously incorrect) since it would have been impossible in practice to assess all information equally. Finally, the aim of the research, as a first exercise in this field, was to investigate *breadth* rather than *depth*. Hence many of the findings are broad, and some may feel that some more detail would have been useful. Where this is the case it is hoped the audit will prove useful in providing a platform for more detailed research by others.

Further information regarding the research methodology employed can be obtained from the Oxford Centre for Disaster Studies, PO Box 137, Oxford, OX4 1BB, UK. Tel: 01865 202772; Fax: 01865 202848.

The International Decade for Natural Disaster Reduction (IDNDR)

Audit Of UK Assets In Disaster Mitigation And Preparedness

Sponsored by the Overseas Development Administration

Questionnaire

to provide information for *The Audit of UK Assets*

Your completion of this questionnaire will provide the information for you/your organisation's entry in the forthcoming publication, *The Audit of UK Assets in Disaster Mitigation and Preparedness.* **You/your organisation will also receive a complementary copy of the audit on its production.**

Please complete and return this questionnaire by **15 February 1995.**

When completed please return to:

David Sanderson, Project Manager,
The Oxford Centre for Disaster Studies (OCDS), PO Box 137, Oxford. OX4 1BB.
If any clarification is required please call 0865 202772 or Fax 0865 202848.

Organisation

1. Name of organisation/individual consultant

2. Address (with post code)

3. Telephone number   4. Fax number

5. E Mail address

6. Contact name and key position (to go as the contact name in *The Audit*)

7. If applicable, please give the names and positions of three other individuals in your organisation with disaster expertise

-
-
-

8. Please indicate if you/your organisation are/is a(n):

☐ Charity/NGO   ☐ Individual Consultant
☐ Private company   ☐ Government Department
☐ Academic/research body   ☐ Intergovernmental agency
☐ Consultancy

Other (please state):

9. Please state when the organisation was formed/you began work in disasters

10. Please briefly state your (organisation's) aims/mission statement

11. Number of full time staff, if any

12. Income for 1993/4   13. Expenditure for 1993/4

Activities

Region
1. Please tick the regions in which you/your organisation works in disaster related activity:

☐	Latin America inc. Mexico	☐	South/South East Asia
☐	Australia and Pacific	☐	East Asia
☐	The Caribbean	☐	Middle East
☐	Western Europe	☐	Africa
☐	Eastern Europe/Former Soviet Union	☐	USA and Canada

Hazard Type
2. Please tick the disaster types in which you/your organisation is involved:

☐	Drought	☐	Earthquake
☐	Hurricane and cyclone	☐	Volcano
☐	Famine	☐	Complex emergency
☐	Flood	☐	Disease and epidemic
☐	Landslide		

Other:

Links
3. Do/does you/your organisation work with southern organisations? Yes/no.

If no, please go to question 4. If yes, please indicate which:

☐	International NGOs	☐	Regional networks
☐	Grassroots NGOs	☐	Government Department
☐	National NGOs	☐	Independent Consultancy
☐	Intergovernmental agency	☐	Academic/research body
☐	Private company		

Other:

4. Regarding your (organisation's) relationship with the southern organisation, please indicate whether you are:

- ☐ Project partner
- ☐ Funder
- ☐ Main contractor
- ☐ Financial resources provider, eg credit loans
- ☐ Training provider
- ☐ Insurance
- ☐ Information exchange

Other:

Skills

5. Please indicate which skills related to disasters you/your organisation includes:

- ☐ Food security
- ☐ Engineering
- ☐ Seismology
- ☐ Building and architecture
- ☐ Physical planning
- ☐ Health/epidemiology/nutrition
- ☐ Agriculture
- ☐ Forestry
- ☐ Hydrology
- ☐ Geomorphology
- ☐ Volcanology
- ☐ Anthropology
- ☐ Conflict prevention/tension reduction
- ☐ Meteorology
- ☐ Insurance/reinsurance
- ☐ AT/indigenous knowledge
- ☐ Energy
- ☐ Remote sensing
- ☐ Development economics
- ☐ Transport
- ☐ Training
- ☐ Information management
- ☐ Technical research
- ☐ Social science research
- ☐ Communications
- ☐ Media

Other:

Work Content

6. Please tick the kinds of work you/your organisation is most closely related to:

- ☐ Community level disaster preparedness
- ☐ Southern public awareness programmes
- ☐ Northern public awareness programme
- ☐ Relief/humanitarian agencies
- ☐ Vulnerability assessment
- ☐ National preparedness planning systems
- ☐ Gender issues
- ☐ Major engineering works
- ☐ Risk assessment
- ☐ Structural mitigation measures
- ☐ Warning systems
- ☐ Goods manufacture
- ☐ Conflict resolution

Other:

Events

7. Do you organise disaster related events? Yes/no.

If no, please go on to question 8. If yes, please indicate which:

- ☐ Workshop
- ☐ Conference
- ☐ Seminar
- ☐ Public exhibitions

Other:

Please list any disaster related **events**, eg conferences, seminars, workshops you/your organisation might be hosting **after June 1995** under the following headings:

Title   Location   Start and finish dates

Current activities
8. Please briefly outline you/your organisation's current disaster related activities under the following headings:

Title   Location and Duration   Brief objectives

Planned initiatives
9. Please outline briefly you/your organisation's planned disaster related initiatives under the following headings:

Title   Location and Duration   Brief objectives

Information

Journals, newsletters, publications

1. Do you/your organisation produce journals, periodicals and/or publications? Yes/no.

If no, go to question 2. If yes, please list any journals and/or periodicals you/your organisation produce(s), under the following headings:

Title	Frequency *eg weekly*	Target audience *eg Planners*	Subscription cost

Please list your (organisation's) most recent disaster related **publications**, if any, under the following headings:

Author(s)	Title	Publisher	Year

Information and Networks

2. Please state the best most recent **'grey literature'** (conference proceedings, reports, discussion papers, etc) you have seen, using where possible the following headings:

Author(s)	Title	Organisation (if any)	Year

3. What resources, if any, do(es) you/your organisation house **which are accessible for outside users**? Please tick:

☐ Library/information centre   ☐ Inquiry service
☐ Photo, video and film library

Other:

4. Please list any professional networks you may belong to:

-
-
-

5. Are you connected to E Mail? Yes/no

If yes, which networks/ bulletin boards do you use? Please list:

-
-
-

6. Please tick any journals/periodicals/newsletters you/your organisation subscribe(s) to:

- ☐ *Hazards Forum*
- ☐ *Disaster Management*
- ☐ *Disasters Journal*
- ☐ *Civil Protection*
- ☐ *Emergency*

Other:

Courses

If you are not involved in training or education, please go to the final question.

1. Please identify any hazard related UK based **training** courses you/your organisation is responsible for, under the following headings:

Title and aims	Duration	Qualification (if any)

2. Please identify any hazard related UK based **academic** courses you/your organisation is responsible for, under the following headings:

Title and aims	Duration	Qualification

Are there any UK based individuals/organisations working in disasters known to you which you think we may have missed? If so, please give

Name   Position   Contact address
                                                         (inc phone/fax if possible)

Do you have any final comments? Please state:

Thankyou for your support

The International Decade for Natural Disaster Reduction (IDNDR)

Audit Of UK Assets In Disaster Mitigation And Preparedness

Sponsored by the Overseas Development Administration

Project Summary

Aim

The aim of the audit is to gain a coherent overview of the current UK assets available in the field of Disaster Mitigation and Preparedness (DMP). It is intended that the audit will provide an up to date resource of information available to the DMP community, funders and interested parties which will:

- Identify strengths, weaknesses and gaps in UK DMP capability (thus enabling the IDNDR Working Group to be more effective)
- Encourage and enhance networking and skillsharing within the DMP community
- Contribute to a more comprehensive and co-ordinated response to international disaster needs.

The audit

The audit, due for publication in Summer 1995, will be organized into the following four categories:

- **Organisations**
  Public, private and charity sector names, addresses and background information
- **Individuals**
  Individual consultants and key individuals in organisations
- **Resources**
  Training/education, financial resources, publications, information and networks, etc
- **Events**
  Forthcoming UK based seminars, workshops and conferences

The audit will be made available in written form and as a database, and it is intended will be updated every 2-3 years.

Project Implementation

The project is implemented on behalf of the Overseas Development Administration by *The Oxford Centre for Disaster Studies* (OCDS) in association with the *Intermediate Technology Development Group* (ITDG).

Any enquiries regarding the audit should be addressed to:

David Sanderson, Project Manager
OCDS, PO Box 137, Oxford. OX4 1BB. Fax 01865 202848.